Diss. ETH No. 30652

Information-Theoretic Aspects of Channel State Quantization

A thesis submitted to attain the degree of
DOCTOR OF SCIENCES
(Dr. sc. ETH Zurich)

presented by

Yiming Yan

M.Sc. ETH
born on February 17, 1996

accepted on the recommendation of
Prof. Dr. Amos Lapidoth, examiner
Prof. Dr. Emre Telatar, co-examiner

2024

Reprint of Diss. ETH No. 30652

ETH Series in Information Theory and its Applications   Vol. 12

edited by Amos Lapidoth

Bibliographic information published by the Deutsche Nationalbibliothek

The Deutsche Nationalbibliothek lists this publication in the Deutsche Nationalbibliografie; detailed bibliographic data are available in the Internet at https://dnb.dnb.de.

First Edition 2024

Hartung-Gorre Verlag Konstanz

ISSN   1860 - 1081
ISBN-10:   3-86628-833-6
ISBN-13:   978-3-86628-833-1

to my family

Acknowledgments

First and foremost, I would like to express my deepest gratitude to my supervisor, Prof. Amos Lapidoth. He introduced me to the fascinating world of information theory as such an amazing teacher, and he granted me the freedom to explore whatever topic that excites me. From him, I learned the essence of insightful and elegant research and the pursuit of precision and clarity in scientific writings. I am also indebted for his constant support and encouragement, which have restored my courage time and again. Working with him has been a true privilege and an unforgettable experience.

I am also deeply grateful to Prof. Emre Telatar for kindly agreeing to be my co-examiner. I feel honored to have a giant in the field as my co-examiner, whose work I have been admiring since my undergraduate years. It has also been a pleasure to chat with him during events and conferences. I would also like to specially thank Danielle Lapidoth-Berger and Dr. Nadia Nibbio, whose encouragement has meant a lot to me. Whichever path I follow, I will remember to keep my shoulder wide.

I feel grateful to have the chance collaborating with Prof. Ligong Wang on the second part of the thesis. His intelligence and humor made every discussion a real pleasure. My thanks extend to Gian Marti, for his valuable comments that eventually triggered an improvement of Chapter 3 and 4 of this thesis. I am also grateful to Prof. Stefan Moser for his help with the German abstract, and for his constant warm and patient response to all my questions—on information theory, on presentations, on LaTeX, on macOS, (and on the coffee machine,) etc. Special gratitude also goes to Prof. Robert Graczyk, who guided me through my first project on Guessing. It turned out to be both adventurous and fruitful, ultimately sparking my pursuit of a Ph.D. degree in information theory. I would also like to thank Prof. Hans-Andrea Loeliger, Prof. Yossef

Steinberg, Prof. Michèle Wigger, Dr. Ran Tamir, and Baohua Ni. The discussions we had in the lab have been a great source of inspiration and joy to me, and I have always been amazed by their brilliance and deep insights.

I am grateful to have spent these four years surrounded by the wonderful colleagues at ISI. Special thanks to Elizabeth (Ruixing), Boxiao, Yiqi, Tianyang, and Yunpeng, for our chats over Chinese dishes, for *Mario Party* and *Pico Park* sessions, and for our badminton and table tennis games at ASVZ. I would also like to thank Patrick, Hampus, Raphael, Guy, Patrik, Reto, Hugo, and Alessio, for the enjoyable time we spent together—in the lab, in the mensa, during the *Bang!* sessions, and along the hiking and skiing routes. Last but not least, I want to thank Olivia Bärtsch and Simone Ammann, for taking care of the administrative work, for maintaining the best working environment, and for always being helpful and caring for our needs.

My sincere gratitude also goes to my friends who made the years in Zurich unforgettable. Special thanks to Yan, Mengfan, Chenyang, and Hao, for our trips across Europe and many adventures; to Tianwei, Ziqing, Chenyu, Yue, Yuqiao, Mafu, Aoming, Zijian, Hanxue, Yuezhu, and Yufeng, for the precious memories from our master's years; to Yueshan, Jie, Farian, Feifei, and Yunni, for our games on the volleyball and badminton courts that have been revitalizing my life; to Shenghao, Zhe, and Lingzheng, for countless chats that have brightened up my days. Their companionship has made this journey both fulfilling and joyful.

I extend my greatest thanks to my beloved family, especially my parents. They have always supported every decision I made. I am profoundly grateful for their unconditional love, endless support, and constant encouragement. I also want to thank them for travelling such a long way from Beijing to Zurich to witness the milestone in my life. This thesis would not have been possible without the support from my family.

Finally, words cannot fully express my gratitude to Yuhang, my best friend and also my family. Together, we screamed on roller coasters in Disneyland and Tivoli; we splashed into seas of Barcelona, Grand Canaria, and Santorini; we witnessed dancing auroras and streaking meteors in the sky; we made a promise to each other in the Stadthaus Zurich... Yet, what I cherish most are the moments we sitting together, chatting over two glasses of wine—sharing our dreams, our childhood, our difficulties and achievements in life, and our thoughts about each other and the rest of the world. Thank you for all of it.

Abstract

This thesis investigates channels with an altruistic helper, a party that observes the channel noise—or more generally, the channel state—and produces rate-limited quantization to assist the data transmission. Various notions of capacity, different revelation of assistance, and the impacts of whether the helper is aware of the message being transmitted or whether a feedback link exists are considered.

The thesis is structured in two primary parts. In the first part, four different notions of capacity are considered on the additive noise channels with a helper:

The first notion, referred to as the *erasures-only capacity*, requires that the decoder must avoid unconscious errors but may declare decoding failures with a small probability. We prove that on the memoryless modulo-additive noise channels (MMANCs), the erasures-only capacity matches the Shannon capacity. This result is generalized to continuous additive noise channels.

The second and third notions considered are the *listsize capacity* and the *cutoff rate*. The listsize capacity requires that the decoder generate a list containing all possible messages, and the ρ-th moment of the cardinality of that list converge to one for given $\rho > 0$. The cutoff rate is similar but restricts the list to messages at least as likely as the transmitted one. It is demonstrated that on the MMANCs, the listsize capacity equals the cutoff rate, and the same result is established on the Gaussian channel with decoder assistance.

The fourth notion examined is the *zero-error capacity*, which requires that the message be decoded with exactly zero probability of error. On the MMANCs, both the scenarios with and without feedback are studied. In its presence, a complete solution of said capacity is provided. In its absence, a solution is provided when the alphabet size is prime. For all other cases, upper and lower bounds on the capacity are derived, leading to a necessary and sufficient condition for its positivity. Thanks to the helper, the zero-error capacity may increase by more than the helper's rate, and it can be positive yet smaller than one bit.

The second part of the thesis focuses on the effects of *message cognizance* and *feedback*:

In particular, the capacity of a state-dependent discrete memoryless channel (SD-DMC) is derived for the setting where a message-cognizant rate-limited helper observes the state sequence noncausally and provides its description to both encoder and decoder. Said capacity is not increased if a feedback link from the receiver to the encoder is introduced.

The same capacity is also derived for the Gaussian channel, and it is demonstrated that message cognizance increases the channel capacity. In this setting the feedback link—while not increasing capacity—eliminates the need for the helper's cognition of the transmitted message. Moreover, in this setting, the results on capacity also hold for the cutoff rate and the listsize capacity.

Keywords: Additive-noise channel; cutoff rate; erasures-only capacity; feedback; helper; listsize capacity; state-dependent channel; zero-error capacity.

Kurzfassung

Diese Dissertation untersucht Kanäle mit einem altruistischen Helfer, der das Kanalrauschen—oder allgemeiner den Kanalzustand—beobachtet und eine ratenbegrenzte Quantisierung zur Unterstützung der Datenübertragung bereitstellt. Dazu werden verschiedene Kapazitätsbegriffe, unterschiedliche Offenlegungen der Unterstützung sowie die Auswirkungen, wenn der Helfer Kenntnis der zu übermittelnden Nachricht hat oder wenn ein Feedback-Link existiert, berücksichtigt.

Die Arbeit ist in zwei Hauptteile gegliedert. Im ersten Teil werden vier verschiedene Kapazitätsbegriffe für additive Rauschkanäle mit einem Helfer betrachtet.

Für den ersten Begriff *Erasures-only-Kapazität* wird verlangt, dass der Dekoder unbewusste Fehler vermeidet, aber es ist ihm erlaubt, mit einer nur kleinen Wahrscheinlichkeit zu deklarieren, dass er nicht dekodieren kann. Wir zeigen, dass auf modulo-additiven Rauschkanälen ohne Gedächtnis (*memoryless modulo-additive noise channels* oder *MMANC*) die Erasures-only-Kapazität gleich ist wie die Shannon-Kapazität. Dieses Resultat wird auf kontinuierliche additive Rauschkanäle verallgemeinert.

Der zweite und der dritte betrachtete Begriff sind die *Listen-Kapazität* und die *Cutoff-Rate*. Für die Listen-Kapazität wird verlangt, dass der Dekoder eine Liste aller möglichen Nachrichten erstellt, wobei das ρ-te Moment der Kardinalität dieser Liste für ein gegebenes $\rho > 0$ gegen eins konvergieren muss. Die Cutoff-Rate ist ähnlich definiert, beschränkt die Liste jedoch auf Nachrichten, die mindestens so wahrscheinlich sind wie die gesendete Nachricht. Es wird gezeigt, dass auf MMANC die Listen-Kapazität und die Cutoff-Rate gleich gross sind und dass dies auch für den Gaußschen Kanal mit Dekoder-Unterstützung gilt.

Der vierte untersuchte Begriff ist die *Null-Fehler-Kapazität*, für die verlangt wird, dass die Nachricht mit einer Fehlerwahrscheinlichkeit von exakt null dekodiert werden kann. Für MMANC untersuchen wir die Situation sowohl mit als auch ohne Feedback-Link. Im Fall mit Feedback leiten wir den exakten Wert der Null-Fehler-Kapazität her. Ohne Feedback präsentieren wir die Null-Fehler-Kapazität in den Situationen, wenn die Grösse des Kanaleingangsalphabets eine Primzahl ist. In den anderen Fällen leiten wir obere und untere Schranken der Null-Fehler-Kapazität her, was zu einer notwendigen und hinreichenden Bedingung für deren Positivität führt. Dank der Unterstützung durch den Helfer kann die Null-Fehler-Kapazität um mehr als die Rate der Unterstützung zunehmen. Zudem ist es möglich, dass sie positiv, aber kleiner als ein Bit ist.

Der zweite Teil der Dissertation konzentriert sich auf die Auswirkungen, wenn der Helfer Kenntnis der zu übermittelnden Nachricht hat oder wenn ein Feedback-Link existiert:

Wir leiten die Kapazität eines zustandsabhängigen Kanals ohne Gedächtnis (*state-dependent discrete memoryless channels* oder *SD-DMC*) her für das Szenario, wenn ein ratenbegrenzter Helfer, der Kenntnis der zu übermittelnden Nachricht hat, den Kanalzustand akausal beobachtet und sowohl dem Enkoder als auch dem Dekoder seine Beschreibung zur Verfügung stellt. Wir zeigen, dass die Kapazität sich nicht erhöht, wenn ein Feedback-Link vom Empfänger zum Sender eingeführt wird. Die gleiche Untersuchung wird auch für den Gaußschen Kanal geführt, und es wird gezeigt, dass sich die Kapazität erhöht, wenn der Helfer Kenntnis der zu übermittelnden Nachricht hat. In diesem Szenario erhöht ein Feedback-Link die Kapazität nicht, aber er eliminiert die Notwendigkeit, dass der Helfer die zu übermittelnde Nachricht kennen muss. Zudem gelten die Ergebnisse zur Kapazität auch für die Cutoff-Rate und die Listen-Kapazität.

Stichworte: Additiver Rauschkanal; Cutoff-Rate; Erasures-only-Kapazität; Feedback-Link; Helfer; Listen-Kapazität; Null-Fehler-Kapazität; zustandsabhängiger Kanal.

Contents

III Appendices   119

A Proofs for Chapter 3   121

B Proofs for Chapter 4   127

C Proofs for Chapter 5   141

D Proofs for Chapter 6   149

Bibliography   155

About the Author   161

Chapter 1

Introduction

This work focuses on channel state quantization and studies its fundamental limits in assisting data transmission from an information-theoretic point of view. In this chapter, we shall explain the motivation for our study, provide a summary of the key topics covered in this thesis, and introduce some notations and definitions that will be used throughout this thesis.

1.1 Motivation

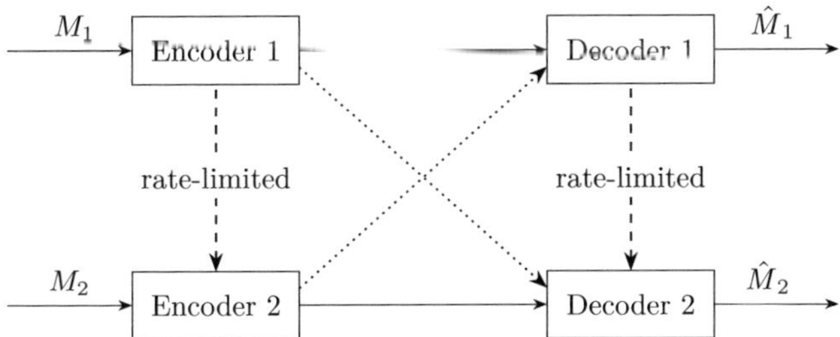

Figure 1.1: Interference channel with cooperation

Consider the interference channel depicted in Fig. 1.1. Two pairs of encoders and decoders communicate over a shared channel, meaning that the transmission from Encoder 1 introduces noise for Decoder 2, and vice

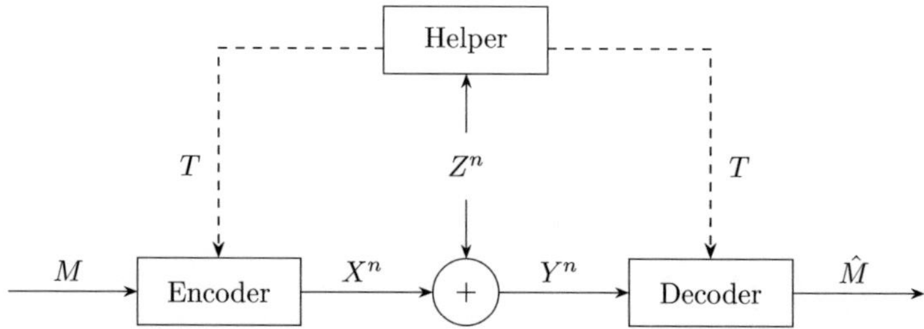

Figure 1.2: Additive noise channels with a helper

versa.

Assume now that users can cooperate, via a rate-limited bit stream sent over a clean and dedicated pipe from User 1 to User 2—to the encoder, to the decoder, or to both. The question is, how to optimally utilize such a rate-limited bit stream to promote User 2's data transmission. Since User 1 knows or can decode the codeword it is transmitting, it has knowledge of the noise encountered by User 2. The challenge then is to determine the best way to quantize this noise to assist data transmission.

Inspired by this, we study the *additive noise channels with a helper*, as illustrated in Fig. 1.2. The channel produces an output as the sum of the input and the noise, with the sum computed either with respect to the real addition in the continuous case, or to the modulo addition when the channel is discrete. A helper is an altruistic party: it observes the channel noise and has the sole target of assisting the data transmission over the channel by producing a rate-limited description of the noise. This description can be made available to the encoder, the decoder, or both, as indicated by the dashed arrows in Fig. 1.2.

For simplicity, assume that the helper has perfect knowledge of the channel noise. We shall study how to optimally perform this quantization, so as to maximize the channel capacity, as a function of the quantization rate. Apart from the multiuser interference channel, this setup is also relevant to applications like interference cancellation for multi-antenna systems, etc.

The additive noise channel with a helper dates back to the work of Bross and Lapidoth [1], where they proved that the channel capacity—the maximum throughput when the probability of decoding error tends to zero—of the Gaussian channel with decoder assistance is increased by the rate of quantization. Later on, this result was generalized to several

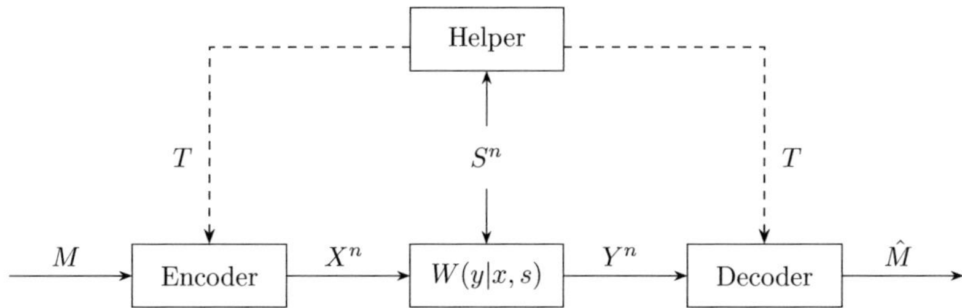

Figure 1.3: State-dependent channels with a helper

multiuser cases [2], and Lapidoth and Gian [3] established the same result when assistance is provided to the encoder, or to both. General stationary and ergodic additive noise channels were studied in [4]. Additionally, the effects of assistance on the error exponent [5], the secrecy capacity [6], the identification capacity [7], and the transmission of a parameter [8], etc., have also been considered. We shall extend these results and study various notions of capacity over the said channel.

Also addressed are the general state-dependent channels, a model of paramount importance in modern communication systems. For example, the channel state information (CSI) on the time-varying fading level of a wireless channel can significantly increase the throughput. Similarly, the CSI in storage devices, representing the physical state of a memory cell before it is written to, is crucial for optimizing data writing processes.

When the CSI is available at the receiver, no matter causally or noncausally, it can be treated as part of the channel output. If available causally at the transmitter, it can be utilized optimally in a symbol-by-symbol manner, where each input depends only on the current state but not the history. The coding is now performed as if over the set of functions mapping each state symbol to an input symbol—this approach is known as Shannon strategy [9]. In cases where the CSI is known noncausally at the transmitter before the entire transmission, Gel'fand and Pinsker [10] solved the channel capacity using the technique of binning.

However, practical constraints often make it challenging to use CSI losslessly. This may occur because the CSI is continuous, or because it is costly or impractical to convey (e.g., via feedback) or utilize (e.g., for encoder design) in a lossless manner. In such scenarios, it becomes essential to quantize the channel state subject to certain rate constraints.

This problem, called *the state-dependent channel with a helper* or *partial CSI*, is illustrated in Fig. 1.3. Dating back to 1983, it was first considered by Heegard and El Gamal [11], where they allowed the encoder and the decoder to receive different descriptions and derived inner and outer bounds on the capacity region. Later on, Rosenzweig, Steinberg, and Shamai solved the capacity when the CSI at the receiver is perfect and that at the transmitter quantized subject to rate constraint [12], while the dual problem, involving perfect noncausal CSI at the transmitter and partial CSI at the receiver is later studied by Steinberg [13]. We refer the readers to [14, Section 3.5] for a comprehensive summary of these results. More recently, Lapidoth and Wang considered a causal helper [15], and Lapidoth and Steinberg further extended the analysis by assuming that the helper is cribbed with past channel inputs [16]. In this thesis, however, the focus is on the noncausal case.

1.2 Outline

This thesis is divided into two parts, each summarized as follows.

1.2.1 Part I: Various Capacities on Additive Noise Channels with a Helper

The first part, corresponding to Chapter 2 to Chapter 5, focuses on the additive noise channels with a helper that—being cognizant of the channel noise sequence noncausally but not of the transmitted message—provides a rate-limited description of said sequence. In this setup, the following notions of capacity will be considered.

Shannon capacity C [17] (or simply *capacity* if clear from the context): It is the maximum achievable rate such that the probability of error in decoding the message vanishes as the blocklength tends to infinity.

Selected results on the Shannon capacity on the additive noise channels with a helper are summarized in Chapter 2, acting as preliminaries for the following parts of the thesis.

Erasures-only capacity $C_{\text{e-o}}$ [18, 19] (a.k.a. *zero-undetected-error capacity* [20] and *zero-error erasure capacity* [21]): Consider a decoder that decodes only if there is a unique message that could have produced the output, and declares an erasure otherwise. $C_{\text{e-o}}$ is the maximum achievable rate for which this probability of erasure tends to zero.

In Chapter 3, we consider the additive noise channels with a helper. On the MMANC, we prove that the erasures-only capacity is identical to the Shannon capacity. This result is generalized to the continuous case.

Order-ρ cutoff rate $\mathsf{R}^{(\rho)}_{\text{cutoff}}$ (a.k.a. the *computational cutoff rate* [22]): Consider a decoder who guesses the transmitted message with a sequence of attempts until a genie confirms that the guess is correct. For $\rho > 0$, $\mathsf{R}^{(\rho)}_{\text{cutoff}}$ is defined as the maximum achievable rate such that the ρ-th moment of the number of guesses approaches one.

This quantity is studied in Chapter 4. It is proved that, on the MMANC, a helper increases the cutoff rate by the rate of assistance (until the channel saturates). This same result is later proved on the Gaussian channel when the assistance is revealed to the decoder.

Order-ρ listsize capacity $\mathsf{C}^{(\rho)}_{\ell}$ [23] (a.k.a. *zero-error list capacity* [24]): Let the decoder produce a list of all the messages that could have produced the output over the channel. For $\rho > 0$, $\mathsf{C}^{(\rho)}_{\ell}$ is the maximum achievable rate with the ρ-th moment of the size of this list approaching one.

The listsize capacity is also studied in Chapter 4, and for all the cases mentioned above, the listsize capacity equals the cutoff rate.

Zero-error capacity C_0: The code is to be decoded without error, so to every output, there can be at most one compatible message that could have produced it. C_0 is then the maximum achievable rate under this condition. It is purely combinatorial, depending only on the support of the channel.

The zero-error capacity on the MMANC with a helper is studied in Chapter 5. It is solved when the cardinality of the alphabet is prime, and for the general case, we derive upper and lower bounds and establish a positivity result for the zero-error capacity.

A noiseless feedback link that provides the encoder with the previously received sequence is also of interest to us, see Fig. 1.4. In particular, we solved all these notions of capacity on the MMANC in the presence of feedback.

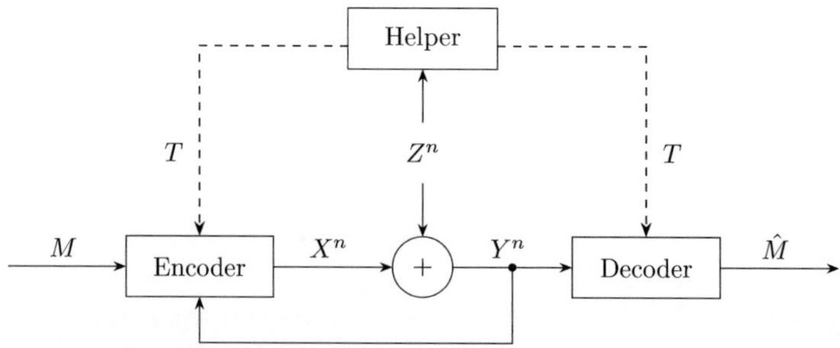

Figure 1.4: Additive noise channels with a helper, in the presence of feedback

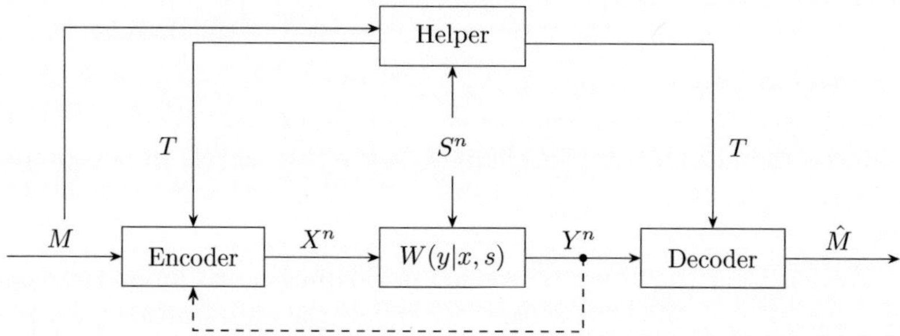

Figure 1.5: State-dependent channels with a message-cognizant bi-terminal helper, in the presence or absence of feedback

1.2.2   Part II: Message Cognizance and Feedback

In the second part, corresponding to Chapter 6, we turn to study the effect of message-cognizance and feedback. In particular, we consider a helper, which observes both the transmitted message and the state sequence noncausally and provides its description to both the encoder and the decoder.

On the SD-DMC, the capacity is derived with above message-cognizant bi-terminal helper, see Fig. 1.5. Even if a feedback link from the receiver to the encoder is introduced, the capacity remains unchanged.

On the Gaussian channel, we derived the same channel capacity, which strictly exceeds that with a message oblivious helper. Now the feedback link does not increase the capacity either, but it eliminates the need for the helper's cognition of the transmitted message. In this sense, the feedback is essentially equivalent as the message cognizance. Moreover, it is demostrated that, in this setup, the cutoff rate and the listsize capacity can be as large as the Shannon capacity.

1.3　Notations

We conclude this chapter with a summary of the notations that are adopted throughout this thesis. Any other notations may be introduced when needed.

Let $\mathcal{X}, \mathcal{Y}$ be finite sets. The cardinality of $\mathcal{X}$ is denoted $|\mathcal{X}|$, and the power set, i.e., the set containing all the subsets of $\mathcal{X}$, is denoted $2^{\mathcal{X}}$. For positive integer n, the n-tuple $(x_1, x_2, \ldots, x_n)$ is denoted x^n or $\mathbf{x}$ (if its length is clear from the context). The set of all n-tuples of elements in $\mathcal{X}$ is denoted $\mathcal{X}^n$.

We use upper case letters, (e.g., X and Y) for chance variables, and lower case letters (e.g., x and y) for their realizations. The set of all probability mass functions (PMFs) on $\mathcal{X}$ is denoted $\mathcal{P}(\mathcal{X})$. If $Q_X \in \mathcal{P}(\mathcal{X})$, we use Q_X^n for the product distribution

$$Q_X^n(x^n) = \prod_{k=1}^{n} Q_X(x_k), \qquad x^n \in \mathcal{X}^n, \tag{1.1}$$

while a generic PMF on $\mathcal{X}^n$ is usually denoted Q_{X^n} or $Q^{(n)}$. If $X^n \sim Q_X^n$, we write $\{X_k\} \sim$ IID Q_X, where IID stands for independent and identically distributed. For $Q_{X,Y} \in \mathcal{P}(\mathcal{X} \times \mathcal{Y})$, we usually use Q_X for the marginal distribution of X and $Q_{X|Y}$ for the conditional distribution of X given Y; we also use $Q_{X|Y=y}$ as the shorthand for $Q_{X|Y}(\cdot|y) \in \mathcal{P}(\mathcal{X})$.

A discrete memoryless channel (DMC) with input alphabet $\mathcal{X}$ and output alphabet $\mathcal{Y}$ is characterized by the transition probability $Q_{Y|X}$. Similarly, $Q_{Y|X}^n$ stands for the n-fold product channel

$$Q_{Y|X}^n(y^n|x^n) = \prod_{k=1}^{n} Q_{Y|X}(y_k|x_k), \qquad x^n \in \mathcal{X}^n, y^n \in \mathcal{Y}^n, \tag{1.2}$$

and a generic channel from $\mathcal{X}^n$ to $\mathcal{Y}^n$ is usually denoted $Q_{Y^n|X^n}$.

For information measures, we follow the notations in [25, Chapter 2]. If chance variables $(X, Y, Z) \sim Q_{X,Y,Z}$, we use $H(X)$ (or $H(Q_X)$) for the entropy of X, $I(X;Y)$ (or $I(Q_X, Q_{Y|X})$) for the mutual information between X and Y, $H(X|Y)$ for the conditional entropy of X given Y, and $I(X;Y|Z)$ for the conditional mutual information between X and Y given Z. For $Q_{X,Y}, P_{X,Y} \in \mathcal{P}(\mathcal{X} \times \mathcal{Y})$, we use $D(P_X \| Q_X)$ for the relative entropy between P_X and Q_X, and $D(P_{X|Y} \| Q_{X|Y} | Q_Y)$ for the conditional relative entropy

$$D(P_{X|Y} \| Q_{X|Y} | Q_Y) = \sum_{y \in \mathcal{Y}} Q_Y(y) D(P_{X|Y=y} \| Q_{X|Y=y}). \qquad (1.3)$$

For $\alpha > 0$, $H_\alpha(X)$ denotes the Rényi entropy of order α, defined as

$$H_\alpha(X) = \frac{\alpha}{1-\alpha} \log \left(\left(\sum_{x \in \mathcal{X}} Q_X(x)^\alpha \right)^{1/\alpha} \right) \qquad (1.4)$$

for $\alpha \in (0,1) \cup (1, +\infty)$ and $H_1(X) = H(X)$; and $H_\alpha(X|Y)$ denotes the conditional Rényi entropy of order α, given by

$$H_\alpha(X|Y) = \frac{\alpha}{1-\alpha} \log \left(\sum_{y \in \mathcal{Y}} \left(\sum_{x \in \mathcal{X}} Q_{X,Y}(x,y)^\alpha \right)^{1/\alpha} \right) \qquad (1.5)$$

for $\alpha \in (0,1) \cup (1, +\infty)$ and $H_1(X|Y) = H(X|Y)$.

We adopt the following basic concepts in the Method of Types, following [25, Section 11.1] and [26, Chapter 3]: Given a length-n sequence $x^n \in \mathcal{X}^n$, its empirical type, or simply type, i.e., the sample frequency of each symbol, is denoted $\hat{P}_{x^n}$. The set of all such types, i.e., all PMFs over $\mathcal{X}$ of denominator n, is denoted $\mathcal{P}_n(\mathcal{X})$. For $P \in \mathcal{P}_n(\mathcal{X})$, the type class of P, i.e., the set of all sequences of type P, is denoted $\mathcal{T}_P^{(n)}$. Analogously, for $x^n \in \mathcal{X}^n$ and $y^n \in \mathcal{Y}^n$, their joint type is denoted $\hat{P}_{x^n, y^n}$, the collection of all joint types denoted $\mathcal{P}_n(\mathcal{X} \times \mathcal{Y})$, and the joint type class for $P \in \mathcal{P}_n(\mathcal{X} \times \mathcal{Y})$ denoted $\mathcal{T}_P^{(n)}$.

Given $\epsilon > 0$ and a PMF $P_X \in \mathcal{P}(\mathcal{X})$, the strongly typical set is denoted $\mathcal{A}_\epsilon^{*(n)}(P_X)$, containing the length-$n$ sequences that are typical with respect to P_X and ϵ as is defined in [25, Section 10.6], and the weakly typical set denoted $\mathcal{A}_\epsilon^{(n)}(P_X)$, defined as in [25, Section 3.1]. For $P_{X,Y} \in \mathcal{P}(\mathcal{X} \times \mathcal{Y})$, the jointly strongly typical set and the jointly weakly typical set with respect to $P_{X,Y}$ are denoted $\mathcal{A}_\epsilon^{*(n)}(P_{X,Y})$ and $\mathcal{A}_\epsilon^{(n)}(P_{X,Y})$, as defined in [25, Section 10.6] and [25, Section 7.6], respectively.

The real numbers are denoted $\mathbb{R}$ and nonnegative reals denoted $\mathbb{R}_{\geq 0}$. Let $\mathcal{B}(\mathbb{R})$ be the Borel σ-field and ν the Lebesgue measure on $\mathbb{R}$. Let

X be a random variable on $(\mathbb{R}, \mathcal{B}(\mathbb{R}))$ of probability measure Q_X. If $Q_X \ll \nu$, we denote its probability density function (PDF, i.e., the Radon–Nykodim derivative with respect to the Lebesgue measure) by f_X, and we write either $X \sim Q_X$ or $X \sim f_X$. As an important example, we say that X is a Gaussian random variable of mean μ and variance σ^2, denoted $X \sim \mathcal{N}(\mu, \sigma^2)$, if

$$f_X(x) = \frac{1}{\sqrt{2\pi\sigma^2}} e^{-\frac{(x-\mu)^2}{2\sigma^2}}, \qquad x \in \mathbb{R}. \tag{1.6}$$

If f_X is the density function of distribution Q_X, we use f_X^n for the density of the product distribution Q_X^n. If random tuple $(X,Y) \sim f_{X,Y}$, we use f_X for the marginal density of X.

Whenever exists, a continuous channel is characterized by the density function $w(\cdot|\cdot)$, with $w(\cdot|x)$ representing the density of the channel output when its input is x. For simplicity of notation, it is extended to vectors as

$$w(y^n|x^n) = \prod_{k=1}^{n} w(y_k|x_k), \qquad x^n, y^n \in \mathbb{R}^n, \tag{1.7}$$

For information measures on continuous random variables, we follow [25, Chapter 8]. If $(X,Y) \sim f_{X,Y}$, we use $h(X)$ (or $h(f_X)$) for the differential entropy of X, $h(X|Y)$ for the conditional differential entropy of X given Y, and $I(X;Y)$ for the mutual information between X and Y.

The weak typicality can be defined for continuous random variables as well. Given $\epsilon > 0$ and a PDF f_X on $\mathbb{R}$, the weakly typical set with respect to f_X is denoted $\mathcal{A}_\epsilon^{(n)}(f_X)$ as [27, Definition 20.23]. Similarly, for a PDF $f_{X,Y}$ on $\mathbb{R}^2$, the jointly weakly typical set with respect to it is denoted $\mathcal{A}_\epsilon^{(n)}(f_{X,Y})$, as is defined in [27, Definition 20.27].

The expectation of a random variable X is denoted $\mathbb{E}[X]$, and the probability of an event E denoted $\mathbb{P}[E]$.

For $\xi, \eta \in \mathcal{A} = \{0, \ldots, |\mathcal{A}| - 1\}$, we use $\xi \oplus \eta$ and $\xi \ominus \eta$ for their mod-$|\mathcal{A}|$ sum and mod-$|\mathcal{A}|$ difference, respectively. They are extended to n-tuples componentwise:

$$\xi^n \oplus \eta^n = \left(\xi_1 \oplus \eta_1, \ldots, \xi_n \oplus \eta_n\right), \qquad \xi^n, \eta^n \in \mathcal{A}^n, \tag{1.8}$$

and likewise for $\xi^n \ominus \eta^n$.

The integers are denoted $\mathbb{Z}$ and positive integers denoted $\mathbb{Z}^+$, and if $n \in \mathbb{Z}^+$, then $[n]$ denotes the set $\{1, 2, \ldots, n\}$. For $\xi \in \mathbb{R}$, we use $\{\xi\}^+$

to denote $\max\{0, \xi\}$, and we use $\lfloor \xi \rfloor$ and $\lceil \xi \rceil$ for its floor (i.e., the largest integer that does not exceed ξ) and ceiling (i.e., the smallest integer that is no less than ξ), respectively. For $\rho > 0$, we often use the abbreviation $\tilde{\rho} \triangleq \frac{1}{1+\rho}$. We use $\mathbb{1}\{\text{statement}\}$ for the indicator function, which equals 1 if the statement is true and 0 otherwise.

Unless stated otherwise, all logarithms are to base 2.

Part I

Various Capacities on Additive Noise Channels with a Helper

Chapter 2

Preliminaries

In the first part of the thesis, we focus on a message-oblivious helper on the additive noise channels—either discrete or continuous—and consider various notions of capacity, i.e., maximal achievable rates under various robustness constraints.

Presented in this chapter are some preliminaries, which will become useful for the following discussions. In Section 2.1, we define the *zero-error list* and the *at-least-as-likely list*, two crucial lists for our discussions on different notions of capacity. Section 2.2 gives a recap on the Shannon capacity of the additive noise channels with a helper.

2.1 Two Crucial Lists

On a general DMC $Q_{Y|X}(\cdot|\cdot)$ with input alphabet $\mathcal{X}$ and output alphabet $\mathcal{Y}$, a blocklength-n code comprises a message set $\mathcal{M} = \{1, 2, \ldots, |\mathcal{M}|\}$ and an encoding function

$$\mathsf{f}\colon \mathcal{M} \to \mathcal{X}^n \tag{2.1a}$$

$$m \mapsto x^n(m) = (x_1(m), \ldots, x_n(m)). \tag{2.1b}$$

Given an output sequence $y^n \in \mathcal{Y}^n$, two lists of messages are crucial: The *zero-error list* (a.k.a. the *remotely-plausible list*)

$$\mathcal{L}_0(y^n) = \big\{ m \in \mathcal{M} \colon Q_{M|Y^n}(m|y^n) > 0 \big\} \tag{2.2}$$

—which contains all messages with positive a posteriori probability—plays a key role in error-free decoding. The *at-least-as-likely list* with respect to the transmitted message m

$$\mathcal{L}(m, y^n) = \{\tilde{m} \in \mathcal{M} : Q_{M|Y^n}(\tilde{m}|y^n) \geq Q_{M|Y^n}(m|y^n)\} \tag{2.3}$$

—which comprises only those messages that are a posteriori at least as likely as m—determines the optimal decoding. Assume, as we do, that the messages are *a priori* equally likely. Then, the a posteriori probability in (2.2) and (2.3) can be replaced equivalently with the likelihood function

$$Q_{Y^n|M}(y^n|m) = \prod_{i=1}^{n} Q_{Y|X}(y_i \,|\, x_i(m)) \tag{2.4}$$

in the sense that they induce identical lists with probability one.

In the presence of a feedback link from the channel output to the encoder, the time-i channel input may depend not only on the message m but also on the past channel outputs y^{i-1}, i.e., a blocklength-n encoder now consists of functions[1]

$$\mathsf{f}_i \colon \mathcal{M} \times \mathcal{Y}^{i-1} \to \mathcal{X} \tag{2.5a}$$
$$(m, y^{i-1}) \mapsto x_i(m, y^{i-1}) \tag{2.5b}$$

for $i \in [n]$. The lists are now defined in the same way but with the likelihood function determined by

$$Q_{Y^n|M}(y^n|m) = \prod_{i=1}^{n} Q_{Y|X}(y_i \,|\, x_i(m, y^{i-1})). \tag{2.6}$$

On the continuous channels, these two lists can be defined identically as in (2.2) and (2.3). If the channel has density $w(\cdot|\cdot)$, then the a posteriori probability $Q_{M|Y^n}$ can be replaced equivalently with the likelihood function, given by

$$f_{Y^n|M}(y^n|m) = \prod_{i=1}^{n} w(y_i \,|\, x_i(m)) \tag{2.7}$$

in the absence of feedback, or

$$f_{Y^n|M}(y^n|m) = \prod_{i=1}^{n} w(y_i \,|\, x_i(m, y^{i-1})) \tag{2.8}$$

---

[1]Also in the presence of feedback, the blocklength n is deterministic: we do not consider coding schemes with random transmission duration.

in its presence.

As we shall see, these lists help us define different notions of capacity and prove relevant results. For instance, the Shannon capacity, which is defined as the maximum achievable rate with the probability of decoding error tending to zero, can be characterized equivalently with the condition

$$\lim_{n\to\infty} \mathbb{P}[|\mathcal{L}(M, Y^n)| \geq 2] = 0, \tag{2.9}$$

because the maximum a posteriori decoding minimizes the probability of error. More examples can be found in Chapter 3 and Chapter 4.

2.2 Channels with a Helper

In this section, we will introduce the channel model, define different types of assistance, and briefly review some previous results on channels with a helper. Some new proofs are also presented.

2.2.1 MMANC

On the MMANC, the time-k output Y_k corresponding to the time-k input x_k is

$$Y_k = x_k \oplus Z_k, \tag{2.10}$$

where x_k, Z_k, and Y_k all take values in the set $\mathcal{A} = \{0, 1, \ldots, |\mathcal{A}| - 1\}$; $\{Z_k\} \sim \text{IID } Q_Z$ is the channel noise with $Q_Z \in \mathcal{P}(\mathcal{A})$. The channel law is thus

$$Q_{Y|X}(y|x) = Q_Z(y \ominus x), \qquad x, y \in \mathcal{A}. \tag{2.11}$$

Recall that "$\oplus$" denotes mod-$|\mathcal{A}|$ addition and "$\ominus$" mod-$|\mathcal{A}|$ subtraction.

Without help, a blocklength-n coding scheme comprises a message set $\mathcal{M}$; an encoder

$$\mathsf{f} \colon \mathcal{M} \to \mathcal{A}^n \tag{2.12a}$$
$$m \mapsto x^n(m); \tag{2.12b}$$

and a generic decoder

$$\phi \colon \mathcal{A}^n \to 2^{\mathcal{M}} \tag{2.13a}$$
$$y^n \mapsto \bar{\mathcal{L}}(y^n), \tag{2.13b}$$

providing either a sole estimate of the transmitted message or a list containing several of them.

A helper, represented by the helping function

$$\mathsf{h} \colon \mathcal{A}^n \to \mathcal{T} \tag{2.14a}$$
$$z^n \to t, \tag{2.14b}$$

is incognizant of the transmitted message M, but observes the noise sequence Z^n noncausally and describes it as T, with T taking values in a finite set $\mathcal{T}$. When presented with assistance (noncausally), an encoder is given by

$$\mathsf{f} \colon \mathcal{M} \times \mathcal{T} \to \mathcal{A}^n \tag{2.15a}$$
$$(m, t) \mapsto x^n(m, t), \tag{2.15b}$$

and a generic decoder given by

$$\phi \colon \mathcal{A}^n \times \mathcal{T} \to 2^{\mathcal{M}} \tag{2.16a}$$
$$(y^n, t) \mapsto \bar{\mathcal{L}}(y^n, t). \tag{2.16b}$$

Depending on to whom the assistance is presented, three types of assistance are considered (see Fig. 1.2):

- *Decoder assistance* corresponds to the scenario where the description T is presented only to the decoder. The encoder is thus of the form (2.12), and the decoder of the form (2.16). Given the output sequence $y^n \in \mathcal{A}^n$ and the helper's description $t \in \mathcal{T}$, the zero-error list $\mathcal{L}_0(y^n, t)$ and the at-least-as-likely list $\mathcal{L}(m, y^n, t)$ are now defined with respect to the likelihood function

$$Q_{Y^n, T|M}(y^n, t|m) = Q_Z^n(y^n \ominus x^n(m)) \cdot \mathbb{1}\{\mathsf{h}(y^n \ominus x^n(m)) = t\}. \tag{2.17}$$

- *Encoder assistance* corresponds to the scenario where T is presented only to the encoder before the entire transmission. The encoder is thus of the form (2.15), and the decoder of the form (2.13). Given y^n, the lists $\mathcal{L}_0(y^n)$ and $\mathcal{L}(m, y^n)$ are now defined with respect to the likelihood function

$$Q_{Y^n|M}(y^n|m) = \sum_{t \in \mathcal{T}} Q_T(t) \, Q_{Z^n|T}(y^n \ominus x^n(m, t)|t). \tag{2.18}$$

- Finally, *bi-terminal assistance* corresponds to the scenario where the same description T is presented to both the encoder and the decoder before the entire transmission. The encoder is thus of the form (2.15), and the decoder of the form (2.16). Given y^n and t, the lists $\mathcal{L}_0(y^n, t)$ and $\mathcal{L}(m, y^n, t)$ can be defined with respect to

$$Q_{Y^n, T|M}(y^n, t|m) = Q_Z^n(y^n \ominus x^n(m, t)) \cdot \mathbb{1}\{h(y^n \ominus x^n(m, t)) = t\}.$$

$$(2.19)$$

Given a sequence of coding schemes, its transmission rate is

$$R = \liminf_{n \to \infty} \frac{1}{n}|\mathcal{M}|, \tag{2.20}$$

and its assistance rate

$$R_h \triangleq \limsup_{n \to \infty} \frac{1}{n}|\mathcal{T}|. \tag{2.21}$$

We emphasize that zero-rate help is not equivalent to no help: it still allows the helper to describe the noise with a sublinear number of bits. This dichotomy does not influence the Shannon capacity, but, as we shall see, is significant for the erasures-only capacity, the listsize capacity, and the zero-error capacity.

The Shannon capacity with rate R_h assistance is defined as the supremum of rate R for which there exists a sequence of coding schemes as above whose probability of error tends to zero, or equivalently,

$$\lim_{n \to \infty} \mathbb{P}[|\mathcal{L}| \geq 2] = 0, \tag{2.22}$$

where $\mathcal{L}$ corresponds to the at-least-as-likely list for each scenario. Using respective subscripts, we denote the Shannon capacity under each type of assistance as a function of R_h by $C_{dec}(R_h)$, $C_{enc}(R_h)$, or $C_{both}(R_h)$. For notational simplicity, we sometimes drop the dependency on R_h when it's clear from the context.

The capacity of the MMANCs with a helper is solved in [2,3]. The results are summarized in the following theorem, the proof of which omitted.

Theorem 2.1 ([2,3]). *On the MMANC with rate R_h assistance,*

$$C_{both}(R_h) = C_{enc}(R_h) = C_{dec}(R_h) \tag{2.23}$$

$$= \log|\mathcal{A}| - \{H(Q_Z) - R_h\}^+. \tag{2.24}$$

In the presence of feedback, the blocklength-n encoder is represented by the sequence of functions

$$\mathsf{f}_i \colon \mathcal{M} \times \mathcal{A}^{i-1} \to \mathcal{A} \tag{2.25a}$$
$$(m, y^{i-1}) \mapsto x_i(m, y^{i-1}), \tag{2.25b}$$

in the absence of assistance, and by

$$\mathsf{f}_i \colon \mathcal{M} \times \mathcal{A}^{i-1} \times \mathcal{T} \to \mathcal{A} \tag{2.26a}$$
$$(m, y^{i-1}, t) \mapsto x_i(m, y^{i-1}, t) \tag{2.26b}$$

when presented with assistance, for $i \in [n]$. The corresponding lists can be constructed by simply replacing $x_i(m)$ or $x_i(m, t)$ in the likelihood function by $x_i(m, y^{i-1})$ or $x_i(m, y^{i-1}, t)$, respectively, for each $i \in [n]$. The capacity with feedback are denoted using subscript by $\mathsf{C}_{\mathrm{F,dec}}(\mathsf{R_h})$, $\mathsf{C}_{\mathrm{F,enc}}(\mathsf{R_h})$, and $\mathsf{C}_{\mathrm{F,both}}(\mathsf{R_h})$.

Remark 2.2. In Chapter 5, as an intermediate result, we establish that feedback does not increase the capacity on the MMANC with a helper (Lemma 5.14). $\triangle$

2.2.2 Continuous Additive Noise Channel

On a general continuous additive noise channel, the time-k output Y_k corresponding to the time-k input x_k is

$$Y_k = x_k + Z_k, \tag{2.27}$$

where x_k, Z_k, and Y_k all take values in the reals and "$+$" stands for real addition. We assume that the channel noise $\{Z_k\} \sim \mathrm{IID}\ Q_Z$ and $Q_Z \ll \nu$, where ν stands for the Lebesgue measure on $\mathbb{R}$. With the PDF of the noise denoted by f_Z, the channel law is thus

$$w(y|x) = f_Z(y - x), \qquad x, y \in \mathbb{R}. \tag{2.28}$$

Without help, a blocklength-n coding scheme comprises a message set $\mathcal{M}$; an encoder

$$\mathsf{f} \colon \mathcal{M} \to \mathbb{R}^n \tag{2.29a}$$
$$m \mapsto x^n(m), \tag{2.29b}$$

such that each codeword $x^n(m)$ satisfies both the average cost constraint

$$\frac{1}{n}\sum_{k=1}^{n} \mathsf{g}(x_k(m)) \leq \mathsf{\Gamma}, \qquad m \in \mathcal{M} \tag{2.30}$$

with $\mathsf{g}\colon \mathbb{R} \to \mathbb{R}_{\geq 0}$ a Borel measurable function and $\mathsf{\Gamma} > 0$, and the amplitude constraint

$$x_k(m) \in [\mathsf{a},\mathsf{b}], \qquad k \in [n],\ m \in \mathcal{M} \tag{2.31}$$

for given $-\infty \leq \mathsf{a} < \mathsf{b} \leq +\infty$ [2]; and a generic decoder

$$\phi\colon \mathbb{R}^n \to 2^{\mathcal{M}} \tag{2.32a}$$
$$y^n \mapsto \bar{\mathcal{L}}(y^n). \tag{2.32b}$$

One important special case is the Gaussian channel: the channel noise is generated IID according to $Q_Z = \mathcal{N}(0,\mathsf{N})$ with $\mathsf{N} > 0$, so

$$w(y|x) = \frac{1}{\sqrt{2\pi\mathsf{N}}}e^{-\frac{(y-x)^2}{2\mathsf{N}}}, \qquad x,y \in \mathbb{R}; \tag{2.33}$$

the encoder satisfies the average power constraint, given by $\mathsf{g}(x) = x^2$ and $\mathsf{\Gamma} - \mathsf{P} > 0$, and the trivial amplitude constraint, given by $\mathsf{a} = -\infty$ and $\mathsf{b} = \infty$.

For simplicity, we further impose the following assumptions.

Assmption 2.3. *The channel and the constraints satisfy:*

- *The noise density f_Z is of finite differential entropy.*

- *$\mathsf{C}^{(\mathrm{I})}$ is finite, where*

$$\mathsf{C}^{(\mathrm{I})} \triangleq \sup_{\substack{Q_X\,:\,\mathbb{E}[\mathsf{g}(X)]\leq\mathsf{\Gamma} \\ Q_X([\mathsf{a},\mathsf{b}])=1}} I(X;Y). \tag{2.34}$$

- *There exists an interval $\mathcal{I}$ of positive Lebesgue measure such that*

$$\mathcal{I} \subseteq [\mathsf{a},\mathsf{b}] \tag{2.35a}$$

   and

$$\mathsf{g}(x) \leq \mathsf{\Gamma}, \qquad x \in \mathcal{I}. \tag{2.35b}$$

---

[2] Here only one average cost constraint and one amplitude constraint are imposed. However, as it becomes clear from the proof, generalization to multiple cost constraints and general support constraint over any $\mathcal{A} \subseteq \mathcal{B}(\mathbb{R})$ is straightforward.

It is known that in this case, the Shannon capacity of the channel is given by [28, Theorem 20.9]

$$\mathsf{C} = \mathsf{C}^{(\mathrm{I})}. \tag{2.36}$$

Also, because the noise is of finite entropy, the Weak Law of Large Numbers (WLLN) guarantees that

$$-\frac{1}{n}\sum_{k=1}^{n} \log f_Z(Z_k) \to \mathbb{E}\big[-\log f_Z(Z)\big] = h(Z) \tag{2.37}$$

in probability, i.e., that the Asymptotic Equipartition Property (AEP) holds. This becomes crucial for Section 3.3 ahead.

A helper, represented by the Borel measurable function

$$\mathsf{h}\colon \mathbb{R}^n \to \mathcal{T} \tag{2.38a}$$
$$z^n \mapsto t, \tag{2.38b}$$

is incognizant of the transmitted message M and describes the noise sequence Z^n as T taking values in finite set $\mathcal{T}$.

When presented with assistance, an encoder is given by

$$\mathsf{f}\colon \mathcal{M} \times \mathcal{T} \to \mathbb{R}^n \tag{2.39a}$$
$$(m,t) \mapsto x^n(m,t), \tag{2.39b}$$

and needs to satisfy the average cost constraint

$$\frac{1}{n}\mathbb{E}\left[\sum_{k=1}^{n} \mathsf{g}(x_k(m,T))\right] \leq \Gamma, \qquad m \in \mathcal{M} \tag{2.40}$$

and the amplitude constraint

$$x_k(m,t) \in [\mathsf{a},\mathsf{b}], \qquad k \in [n],\ m \in \mathcal{M},\ t \in \mathcal{T}; \tag{2.41}$$

and a decoder is given by

$$\phi\colon \mathbb{R}^n \times \mathcal{T} \to \mathcal{M} \tag{2.42a}$$
$$(y^n,t) \mapsto \bar{\mathcal{L}}(y^n,t). \tag{2.42b}$$

Three different assistance revelation are considered: *decoder assistance, encoder assistance,* and *bi-terminal assistance*; the zero-error list $\mathcal{L}_0$ and the at-least-as-likely list $\mathcal{L}$ can be constructed as in Section 2.2.1 by replacing Q_Z by the density function f_Z, and the Shannon capacity is defined accordingly. Feedback can be introduced in the same way as in Section 2.2.1.

The study of the Gaussian channel with a helper dates back to [1]. Later on, it is generalized to any stationary and ergodic noise process of finite variance, under general average cost and support constraint [4]. More discussion on the decoder-assisted memoryless additive noise networks under average-power input constraints can be found in [2]. Lower bounds to the capacity of general additive noise channels with encoder or bi-terminal assistance were proposed [3], which are known to be tight on the Gaussian and the Exponential channels. For the general noise, the capacity remains open. Some selected results are summarized in the following theorem.

Theorem 2.4 ([1]–[4]). *On the general memoryless additive noise channel subject to Assumption 2.3,*

$$\mathsf{C}_{\mathrm{both}}(\mathsf{R}_{\mathrm{h}}) \geq \mathsf{C}_{\mathrm{enc}}(\mathsf{R}_{\mathrm{h}}) \geq \mathsf{C}_{\mathrm{dec}}(\mathsf{R}_{\mathrm{h}}) = \mathsf{C} + \mathsf{R}_{\mathrm{h}}, \qquad (2.43)$$

where C is the capacity without help as in (2.36). The inequalities hold with equality on the Gaussian channel with second-moment constraint.

Remark 2.5. The assumptions in Theorem 2.4 are different from those in the previous work: For the decoder assistance, it was assumed that the noise is of finite second moment [2, 4]. For the encoder assistance, as in [3, Theorem 3], assumptions on the noise include that $\mathbb{E}\left[Z^{2+\delta}\right]$ is finite for some $\delta > 0$, and that

$$\|f_Z\|_{1/3} = \left(\int_{\mathbb{R}} f_Z(z)^{1/3}\,\mathrm{d}z\right)^3 < \infty; \qquad (2.44)$$

and the encoder is subject to the average power constraint. $\qquad \triangle$

A new proof of Theorem 2.4, which adopts a modulo quantization scheme, is presented in the following subsection. On the one hand, it simplifies the original proofs and lessens the technical conditions as indicated by Remark 2.5. On the other hand, it enhances the result by proving that the extra R_{h} bits per channel use can be conveyed *error-free* with negligible overhead, which becomes a key building block for this thesis.

Remark 2.6. As we shall see in the proof, the coding scheme is in fact based on *causal help*—This is also the case for Section 3.3 ahead. Since causality is not the main concern of this thesis, we refrain from a more detailed discussion. $\triangle$

2.2.3 Proof of Theorem 2.4

We first focus on the decoder assistance, and the converse bound follows from the Cut-Set Bound [25, Theorem 15.10.1].

The key to proving the achievability of $C + R_h$ is in showing that rate-R_h help can be utilized to increase the data rate by R_h, and indeed, this can be done error-free with negligible overhead. To that end, we show that—by sending over the channel a single input symbol that takes value in an interval $\mathcal{I}$ subject to (2.35)—we can convey a message taking values in $\{0, \ldots, \kappa - 1\}$ error-free with the assistance taking values in the same set.

Without loss of generality, let $\mathcal{I} = [0, \alpha]$ with $\alpha > 0$ (otherwise shift both the interval and the constraints by a constant). To transmit $m \in \{0, \ldots, \kappa - 1\}$, the encoder sends

$$x = m \cdot \frac{\alpha}{\kappa} \in \mathcal{I}. \tag{2.45}$$

Upon observing the noise Z, the helper produces the description T by quantizing the normalized noise and taking modulo, i.e.,

$$T = \left\lfloor Z \cdot \frac{\kappa}{\alpha} \right\rfloor \mod \kappa, \tag{2.46}$$

which is an element of $\{0, \ldots, \kappa - 1\}$. Based on Y and T, the decoder can calculate

$$\hat{m} = \left\lfloor Y \cdot \frac{\kappa}{\alpha} - T \right\rfloor \mod \kappa, \tag{2.47}$$

which equals m with probability one, because

$$\hat{m} = \left\lfloor (x + Z) \cdot \frac{\kappa}{\alpha} - T \right\rfloor \mod \kappa \tag{2.48}$$

$$= \left\lfloor m + Z \cdot \frac{\kappa}{\alpha} - T \right\rfloor \mod \kappa \tag{2.49}$$

$$= \left(m + \left\lfloor Z \cdot \frac{\kappa}{\alpha} \right\rfloor - T \right) \mod \kappa \tag{2.50}$$

$$= m, \tag{2.51}$$

where (2.50) holds because m and T are both integers.

Using this building block, we can now prove the achievability of $\mathsf{C}+\mathsf{R}_\mathsf{h}$ following the idea of "flash-helping" [2]–[4], where the noise description is provided infrequently but with great precision. Specifically, we propose the following blocklength-$(n+1)$ scheme: In Phase 1, we use channel once with the aforementioned coding scheme by letting $\kappa = \lceil 2^{n\mathsf{R}_\mathsf{h}} \rceil$. In Phase 2 of length n, the channel is used without help at transmission rate $\mathsf{R} < \mathsf{C}$.

It is easy to verify that the input sequence satisfies

$$\sum_{k=1}^{n+1} \mathsf{g}(x_k) = \mathsf{g}(x_1) + \left(\sum_{k=2}^{n+1} \mathsf{g}(x_k) \right) \leq (n+1)\Gamma, \qquad (2.52)$$

and

$$x_k \in [\mathsf{a}, \mathsf{b}], \qquad k \in [n+1]. \qquad (2.53)$$

The overall transmission rate is

$$\lim_{n\to\infty} \frac{1}{n+1} \left(\log\lceil 2^{n\mathsf{R}_\mathsf{h}} \rceil + n\mathsf{R} \right) = \mathsf{R} + \mathsf{R}_\mathsf{h}, \qquad (2.54)$$

and the rate of help is

$$\lim_{n\to\infty} \frac{1}{n+1} \left(\log\lceil 2^{n\mathsf{R}_\mathsf{h}} \rceil + 0 \right) = \mathsf{R}_\mathsf{h}. \qquad (2.55)$$

The achievability now follows by noting that, since Phase 1 is error-free, the overall probability of error equals that in Phase 2, which can be made arbitrarily small whenever $\mathsf{R} < \mathsf{C}$. We thus conclude that

$$\mathsf{C}_{\mathrm{dec}}(\mathsf{R}_\mathsf{h}) = \mathsf{C} + \mathsf{R}_\mathsf{h}. \qquad (2.56)$$

In the presence of encoder assistance, the achievability is similar. Except that to transmit a message taking values in $\mathcal{T} = \{0, \ldots, \kappa - 1\}$ error-free in the first channel use with the help of

$$T = \left\lfloor Z \cdot \frac{\kappa}{\alpha} \right\rfloor \mod \kappa, \qquad (2.57)$$

the encoder now pre-subtracts T from the message and sends

$$x = \left((m - T) \mod \kappa \right) \cdot \frac{\alpha}{\kappa} \in \mathcal{I}. \qquad (2.58)$$

The decoder produces

$$\hat{m} = \left\lfloor Y \cdot \frac{\kappa}{\alpha} \right\rfloor \mod \kappa, \tag{2.59}$$

which equals m with probability one, because

$$\hat{m} = \left\lfloor (x + Z) \cdot \frac{\kappa}{\alpha} \right\rfloor \mod \kappa \tag{2.60}$$

$$= \left\lfloor \left((m - T) \mod \kappa \right) + Z \cdot \frac{\kappa}{\alpha} \right\rfloor \mod \kappa \tag{2.61}$$

$$= \left(m - T + \left\lfloor Z \cdot \frac{\kappa}{\alpha} \right\rfloor \right) \mod \kappa \tag{2.62}$$

$$= m. \tag{2.63}$$

The achievability of any rate up to $\mathsf{C} + \mathsf{R_h}$ now follows from "flash-helping" in the same way as decoder assistance.

Finally, we conclude the proof by noting that $\mathsf{C_{both}}(\mathsf{R_h}) \geq \mathsf{C_{enc}}(\mathsf{R_h})$ because assistance never hurts. $\qquad\square$

Chapter 3

Erasures-Only Capacity with a Helper

3.1   Introduction

In this chapter, the erasures-only capacity is computed for the MMANC with a helper, who provides a rate-limited description of the noise sequence to the decoder, to the encoder, or to both. The result is generalized to continuous additive noise channels. In particular, the erasures-only capacity on the Gaussian channel with a helper equals the Shannon capacity, while for the general case, a lower bound is derived. In all scenarios the gains in these capacities thanks to the helper can exceed the helper's rate.

In the remaining part of this section, we will define erasures-only capacity and discuss some of its properties. In Section 3.2, we present the results on the MMANC with a helper; and in Section 3.3, these results will be generalized to the continuous case.

Consider a general DMC $Q_{Y|X}$ for now. An erasures-only decoder is not allowed to make an unconscious mistake but can declare an erasure in case of uncertainty, claiming a failure in decoding. The erasures-only capacity $C_{\text{e-o}}$ is defined as the supremum of achievable rates such that the probability of erasure tends to zero.

Note that the decoder must declare an erasure if the zero-error list $\mathcal{L}_0$

contains more than one message, and otherwise produce the sole message in the list. Thus, $\mathsf{C}_{\text{e-o}}$ can be equivalently defined as the supremum of achievable rates under the condition

$$\lim_{n \to \infty} \mathbb{P}\big[|\mathcal{L}_0| \geq 2\big] = 0. \tag{3.1}$$

Because Shannon capacity is defined by replacing $\mathcal{L}_0$ with $\mathcal{L}$ in (3.1), and $\mathcal{L} \subseteq \mathcal{L}_0$ with probability one, we conclude that

$$\mathsf{C}_{\text{e-o}} \leq \mathsf{C}. \tag{3.2}$$

The erasures-only capacity is neither continuous in the channel matrix (like the Shannon capacity) nor purely combinatorial (like the zero-error capacity). Results on the erasures-only capacity of general DMCs are scarce. Noteworthy exceptions are the results of Pinsker and Sheverdjaev [18] and Csiszár and Narayan [19] that provide sufficient conditions for the erasures-only capacity to equal the Shannon capacity. Asymptotic results on the erasures-only capacity in the low-noise regime can be found in [21] and [20].

A single-letter expression for the erasures-only capacity remains open, while a multi-letter characterization was proved by Ahlswede et al. [21] and Telatar [29, 30] as

$$\mathsf{C}_{\text{e-o}} = \lim_{k \to \infty} \frac{1}{m} \max_{P^{(m)}} \inf_{\substack{\hat{W}^{(m)} \ll Q^m_{Y|X} : \\ P^{(m)} W^{(m)} = P^{(m)} Q^m_{Y|X}}} I\big(P^{(m)}, Q^m_{Y|X}\big), \tag{3.3}$$

where $P^{(m)}$ ranges over the probability distributions on $\mathcal{X}^m$, and $\hat{W}^{(m)}$ over the transition probabilities from $\mathcal{X}^m$ to $\mathcal{Y}^m$; the notion $\hat{W} \ll W$ means that $\hat{W}(y|x) = 0$ whenever $W(y|x) = 0$; and PW represents the output distribution of the channel W when the input distribution is P.

Even for some simplest channels, the erasures-only capacity remains unknown. For example, Bunte et al. [31] considered the cyclic triangle, i.e., the MMANC with $|\mathcal{A}| = 3$ and $Q_Z(1) = 1 - Q_Z(0) = \epsilon$ for $1 > \epsilon > 0$. They only established that the erasures-only capacity is upper bounded by $\log 2$, and approaches $\log 2$ when $\epsilon \downarrow 0$.

In the presence of feedback, the erasures-only capacity is determined by Bunte et al. [32]: The erasures-only capacity with feedback $\mathsf{C}_{\text{e-o,F}}$ is zero if $\mathsf{C}_{\text{e-o}} = 0$, and equals the Shannon capacity otherwise.

3.2 MMANC

Now we consider the MMANCs with a rate-R_h helper. The erasures-only capacity under each type of assistance is defined analogously as the supremum of R, for which there exists a sequence of such coding schemes of rate-R under rate-R_h help, subject to (3.1). We establish the following result, namely, the helper raises the erasures-only capacity to the same value to which it raises the Shannon capacity, irrespective of whom the description is revealed to.

Theorem 3.1. *The erasures-only capacity of the MMANC with rate-R_h assistance matches the Shannon capacity, i.e., is given by*

$$\mathsf{C}_{\text{e-o,both}}(\mathsf{R}_h) = \mathsf{C}_{\text{e-o,enc}}(\mathsf{R}_h) = \mathsf{C}_{\text{e-o,dec}}(\mathsf{R}_h) \tag{3.4}$$

$$= \log|\mathcal{A}| - \big\{H(Q_Z) - \mathsf{R}_h\big\}^+. \tag{3.5}$$

Remark 3.2. The results when the rate of help is zero may seem paradoxical: even if zero in the absence of help, the erasures-only capacity with zero-rate help is equal to the Shannon capacity. This paradox is resolved by noting that zero-rate help is not equivalent to no help: Zero-rate help still allows the helper to describe the noise, albeit with a number of bits that is subexponential in the blocklength. As we shall see, such a description is all it takes to raise the erasures-only capacity to the Shannon capacity. $\triangle$

As a corollary, feedback does not increase the erasures-only capacity with help.

Corollary 3.3. *The erasures-only capacity of the MMANC with rate-R_h assistance in the presence of feedback is given by*

$$\mathsf{C}_{\text{e-o,both,F}}(\mathsf{R}_h) = \mathsf{C}_{\text{e-o,enc,F}}(\mathsf{R}_h) = \mathsf{C}_{\text{e-o,dec,F}}(\mathsf{R}_h) \tag{3.6}$$

$$= \log|\mathcal{A}| - \big\{H(Q_Z) - \mathsf{R}_h\big\}^+. \tag{3.7}$$

Proof. The direct part follows from Theorem 3.1, and the converse part from the relation (3.2) and the fact that feedback does not increase the Shannon capacity with help on the MMANC (Remark 2.2). $\square$

In the following subsection, we prove Theorem 3.1.

3.2.1 Proof of Theorem 3.1

The right-hand side (RHS) of (3.5) is the Shannon capacity of this channel with rate-R_h assistance (Theorem 2.1). Since the erasures-only capacity never exceeds the Shannon capacity, the converse follows. Thus, only the achievability results for $C_{e\text{-}o,dec}$ and $C_{e\text{-}o,enc}$ need to be established—that for $C_{e\text{-}o,both}$ then follows from the fact that assistance never hurts. We first focus on the decoder assistance, on which the proof for encoder assistance will be built. Consider three different cases:

- Case 1: $R_h = 0$.

Let Q_X be the uniform input distribution, $Q_{X,Y}$ the joint input-output distribution it induces, and Q_Y the corresponding (uniform) output distribution. Fix $\epsilon > 0$. Generate a random codebook $\{X^n(m)\}_{m=1,\ldots,2^{nR}}$ of independent codewords, each having components drawn IID from Q_X.

To send the message $m \in \mathcal{M}$, the encoder transmits the m-th codeword. The helper produces a one-bit description $T = \mathbb{1}\{Z^n \in \mathcal{A}_\epsilon^{(n)}(Q_Z)\}$ of the noise sequence indicating whether or not it is weakly typical. The decoder receives the tuple (Y^n, T). If $T = 0$, it declares an erasure; otherwise, it searches for a message $\tilde{m}$ such that $(X^n(\tilde{m}), Y^n) \in \mathcal{A}_\epsilon^{(n)}(Q_{X,Y})$. If such an $\tilde{m}$ exists and is unique, it produces $\tilde{m}$. Otherwise, it declares an erasure.

To prove that undetected errors never occur, we note that the decoder attempts to decode only if $T = 1$, and that, as we next show, in this case, the transmitted codeword $X^n(m)$ is jointly typical with the received sequence Y^n. Indeed, since Q_X and Q_Y are equiprobable, it follows that for all $x^n, y^n \in \mathcal{A}^n$

$$-\frac{1}{n}\log Q_X^n(x^n) = H(Q_X) \quad \text{and} \quad -\frac{1}{n}\log Q_Y^n(y^n) = H(Q_Y), \quad (3.8)$$

and, consequently, $X^n(m)$ is jointly typical with Y^n because

$$\left| -\frac{1}{n}\log Q_{X,Y}^n(X^n(m), Y^n) - H(X,Y) \right|$$

$$= \left| -\frac{1}{n}\log Q_{Y|X}^n(Y^n|X^n(m)) - H(Y|X) \right| \quad (3.9)$$

$$= \left| -\frac{1}{n}\log Q_Z^n(Z^n) - H(Z) \right| \quad (3.10)$$

$$\leq \epsilon, \quad (3.11)$$

where (3.9) follows from the chain rule and (3.8); (3.10) holds because $Y = X \oplus Z$, with Z independent of X; and (3.11) holds because, when T is 1, $Z^n \in \mathcal{A}_\epsilon^{(n)}(Q_Z)$.

It remains to establish that the probability of erasure vanishes as n tends to infinity. An erasure is declared only if Z^n is atypical or if $(X^n(\tilde{m}), Y^n) \in \mathcal{A}_\epsilon^{(n)}(Q_{X,Y})$ for some $\tilde{m} \neq m$. The probability of the former tends to zero by the AEP [25], and the probability of the latter tends to zero whenever

$$\mathsf{R} < I(Q_X, Q_{Y|X}) = \log|\mathcal{A}| - H(Q_Z) \tag{3.12}$$

and ϵ is sufficiently small.

- Case 2: $\mathsf{R}_\mathrm{h} > H(Q_Z)$.

  Fix $0 < \epsilon < \mathsf{R}_\mathrm{h} - H(Q_Z)$. The codebook we use in this case comprises all the distinct sequences in $\mathcal{A}^n$. The helper indicates by the bit $T_1 = 1/0$ whether the noise sequence is typical/atypical. If it is typical, the helper provides an almost lossless description of the noise sequence by producing as T_2 its index in $\mathcal{A}_\epsilon^{(n)}(Q_Z)$. (Otherwise, T_2 is arbitrary.) The decoder, upon receiving $(Y^n, (T_1, T_2))$, declares an erasure if the noise is atypical, as indicated by $T_1 = 0$. Otherwise, it reconstructs the noise sequence from T_2 and subtracts it from Y^n to recover the codeword and hence the message. The rate $\log|\mathcal{A}|$ is thus achievable.

- Case 3: $0 < \mathsf{R}_\mathrm{h} \leq H(Q_Z)$.

  For any $\delta > 0$, we divide the transmission block into two parts of relative length $\frac{\mathsf{R}_\mathrm{h}}{(1+\delta)H(Q_Z)}$ and $1 - \frac{\mathsf{R}_\mathrm{h}}{(1+\delta)H(Q_Z)}$. We then apply the aforementioned coding schemes for helper rates of $(1+\delta)H(Q_Z)$ and zero, respectively. By the Union Bound, the overall probability of undetected errors is zero, and that of erasures tends to zero. Thus, the total rate achieved by this time-sharing scheme is

$$\frac{\mathsf{R}_\mathrm{h} \log|\mathcal{A}|}{(1+\delta)H(Q_Z)} + \left(1 - \frac{\mathsf{R}_\mathrm{h}}{(1+\delta)H(Q_Z)}\right)\left(\log|\mathcal{A}| - H(Q_Z)\right)$$
$$= \log|\mathcal{A}| - H(Q_Z) + \frac{\mathsf{R}_\mathrm{h}}{1+\delta}. \tag{3.13}$$

The result follows by taking $\delta \downarrow 0$. This establishes the direct part for decoder assistance.

It remains to prove the achievability for the encoder-assisted case. The idea is to convey the assistance to the decoder with negligible loss in transmission rate. We similarly consider three cases:

- Case 1: $R_h = 0$.

We propose the following blocklength-$(n + 1)$ scheme. For any $R < \log|\mathcal{A}| - H(Q_Z)$ and $\epsilon > 0$ sufficiently small, consider a rate-R blocklength-n codebook that, when used with decoder assistance as is proposed above, yields no undetected errors and a small probability of erasure. Using $(1 + \lceil \log|\mathcal{A}| \rceil)$ bits, the helper conveys to the encoder the bit $T_1 = \mathbb{1}\{Z^n \in \mathcal{A}_\epsilon^{(n)}(Q_Z)\}$ and the noise sample $T_2 = Z_{n+1}$. To send the message m, the encoder transmits the codeword $x^n(m)$, followed by $X_{n+1} = T_1 \ominus T_2$, so that $Y_{n+1} = T_1$. Equipped with T_1, the decoder can proceed as if with decoder assistance. The overall rate is $nR/(n + 1)$, which approaches R.

- Case 2: $R_h > H(Q_Z)$.

Fix $0 < \epsilon < R_h - H(Q_Z)$. As in Case 1, the helper and encoder can convey the bit $T_1 = \mathbb{1}\{Z_1^n \in \mathcal{A}_\epsilon^{(n)}(Q_Z)\}$ to the decoder. Additionally, if the noise is typical, the helper can describe it to the encoder, who can then subtract it from the codeword.

- Case 3: $0 < R_h \leq H(Q_Z)$.

Follows by time sharing similarly. $\square$

3.3   Continuous Additive-Noise Channels

On general cost-constrained additive-noise channels, the erasures-only capacity under each type of assistance is defined subject to (3.1), analogously as in Section 3.2.

Without help, the erasures-only capacity is zero on the Gaussian channel: Because f_Z is supported everywhere, no codeword is ruled out by any received sequence, and an erasure must be declared with probability one. However, as we shall see in the following theorem, a little bit of help makes a huge difference.

Theorem 3.4. *For general additive-noise channels subject to Assumption 2.3 and with rate-R_h decoder assistance, the erasures-only capacity is identical to the Shannon capacity, i.e., is given by*

$$C_{\text{e-o,dec}}(R_h) = C + R_h, \tag{3.14}$$

where C *is the channel capacity without help given by*

$$\mathsf{C} = \sup_{\substack{Q_X \,:\, \mathbb{E}[\mathsf{g}(X)] \leq \mathsf{\Gamma} \\ Q_X([\mathsf{a},\mathsf{b}])=1}} I(X;Y). \qquad (3.15)$$

Theorem 3.5. *For general additive-noise channels subject to Assumption 2.3 and with the rate-R_h encoder or bi-terminal assistance,*

$$\mathsf{C}_{\mathrm{e\text{-}o,both}}(\mathsf{R}_\mathrm{h}) \geq \mathsf{C}_{\mathrm{e\text{-}o,enc}}(\mathsf{R}_\mathrm{h}) \geq \mathsf{C} + \mathsf{R}_\mathrm{h}. \qquad (3.16)$$

where C *is the channel capacity without help. Both inequalities hold with equality for the Gaussian channel, i.e., when f_Z is Gaussian, g quadratic, $\mathsf{a} = -\infty$, and $\mathsf{b} = +\infty$.*

The proofs of Theorem 3.4 and Theorem 3.5 are presented in the following subsections.

Remark 3.6. One may wonder, how much generality we lose by assuming that Q_Z is absolutely continuous with respect to the Lebesgue measure. We give a partial answer to this question in the following proposition, whose proof is discussed in Appendix A.1.

Proposition 3.7. *In continuous additive noise channels whose input constraint satisfies (2.35), if the noise distribution Q_Z contains point mass, then the erasures-only capacity already equals infinity even without helper—a fortiori, it is infinity in the presence of assistance.*

$\triangle$

3.3.1   Proof of Theorem 3.4

The converse holds since

$$\mathsf{C}_{\mathrm{e\text{-}o,dec}}(\mathsf{R}_\mathrm{h}) \leq \mathsf{C}_{\mathrm{dec}}(\mathsf{R}_\mathrm{h}) = \mathsf{C} + \mathsf{R}_\mathrm{h}, \qquad (3.17)$$

where the second inequality follows from Theorem 2.4.

The achievability follows from a time-sharing argument. With the scheme in Section 2.2.3, using only input symbols in $\mathcal{I}$, within one channel use and $\lceil n\mathsf{R}_\mathrm{h}\rceil$-bit description of the noise, $\lceil n\mathsf{R}_\mathrm{h}\rceil$ bits can be transmitted *error-free.* So we only need to prove the achievability of rate C with zero-rate help, or more precisely—as we shall see—with one bit of help.

Fix $\mathsf{R}, \epsilon > 0$ to be specified later, and let $\mathcal{M} = \{1, \ldots, 2^{n\mathsf{R}}\}$. We consider an auxiliary setup, where instead of the erasures-only decoder, we insist on using a modified threshold decoder ϕ_{Th} (without help), which produces $\phi_{\mathrm{Th}}(y^n) = m$ if the corresponding codeword satisfies

$$\frac{1}{n} \sum_{k=1}^{n} \mathsf{g}(x_k(m)) \leq \Gamma \tag{3.18}$$

and

$$x_k(m) \in [\mathsf{a}, \mathsf{b}], \qquad k \in [n], \tag{3.19}$$

and the likelihood function (which in our case is $w^n(y^n|x^n) = f_Z^n(y^n - x^n)$) satisfies

$$f_Z^n(y^n - x^n(m)) > 2^{n\epsilon} f_Z^n(y^n - x^n(m')), \qquad m' \in \mathcal{M} \setminus \{m\}; \tag{3.20}$$

if no such m exists, it produces $\phi_{\mathrm{Th}}(y^n) = 0$. We claim the following.

Claim 3.8. *Consider an additive-noise channel of noise density f_Z, whose capacity*

$$\mathsf{C} = \sup_{\substack{Q_X \,:\, \mathbb{E}[\mathsf{g}(X)] \leq \Gamma \\ Q_X([\mathsf{a},\mathsf{b}]) = 1}} I(X; Y) \tag{3.21}$$

is finite. Then, for any $\epsilon > 0$, there exists a sequence of codebooks of rate up to $\mathsf{C} - 4\epsilon$, such that the maximal *probability of decoding error under ϕ_{Th} vanishes as the blocklength tends to infinity.*

The existence of such a sequence essentially follows from [33, Lemma 2], and the details are postponed to Appendix A.2. We proceed to denote these codebooks (indexed by n) by $\mathcal{C}_n = \{x^n(1), \ldots, x^n(|\mathcal{M}|)\}$. Note that, *a fortiori*, every codeword in $\mathcal{C}_n$ satisfies the input constraints.

Now we propose our erasures-only coding scheme with one bit of help. To transmit $m \in \mathcal{M}$, the encoder sends $x^n(m)$. The helper, upon observing Z^n, produces a one-bit description

$$T = \mathbb{1}\{Z^n \in \mathcal{A}_{\epsilon/2}^{(n)}(f_Z)\}, \tag{3.22}$$

indicating whether or not the noise sequence is weakly typical with constant $\epsilon/2$. The decoder, upon observing Y^n and T, performs the following:

- If $T = 0$, it declares an erasure;

- If $T = 1$, it seeks some $m' \in \mathcal{M}$ such that

$$Y^n - x^n(m') \in \mathcal{A}_{\epsilon/2}^{(n)}(f_Z), \tag{3.23}$$

and if there is a unique such m', it returns $\hat{m} = m'$ as its guess; else, it declares an erasure.

We next analyze the probability of erasure and undetected error.

It is easy to see that an undetected error never occurs: The decoder attempts to decode only if $T = 1$, which guarantees that $Y^n - x^n(m)$ is typical for the transmitted message m. So either there exists $m' \neq m$ that happens to meet the criteria—which leads to an erasure—or m is the unique message subject to (3.23) and is decoded correctly.

As for the probability of erasure, note that this happens either when $T = 0$—which has vanishing probability by AEP—or when there exists $m' \neq m$ that also meets the criterion (3.23). The probability of the latter is vanishing because this event leads to an error under ϕ_{Th}: Indeed, (3.23) being true for both m and m' implies

$$-\frac{1}{n} \log f_Z^n(Y^n - x^n(m')) \leq -\frac{1}{n} \log f_Z^n(Y^n - x^n(m)) + \epsilon, \tag{3.24}$$

and thus, $\phi_{\mathrm{Th}}(Y^n) \neq m$. This establishes the achievability of rate R, and the theorem follows by letting $\epsilon \to 0$. $\qquad\square$

3.3.2 Proof of Theorem 3.5

We propose the following blocklength-$(n + 2)$ scheme. Phase 1 is of length 1, Phase 2 of length n, and Phase 3 of length 1.

In Phase 1, we adopt the scheme in Section 2.2.3: The helper produces T_1 of $\lceil n\mathsf{R_h} \rceil$ bits as a quantization of Z_1, and the encoder can thus transmit $\lceil n\mathsf{R_h} \rceil$ bits of messages error-free within one channel use and with negligible overhead. In Phase 2, similar to the decoder-assisted scheme as in Section 3.3.1, the helper produces T_2 of one bit reflecting whether or not Z_2^{n+1} is typical, which will be conveyed to the decoder error free in Phase 3. With this knowledge at the decoder, Phase 2 reduces to decoder-assisted transmission, and any rate $\mathsf{R} < \mathsf{C}$ is achievable due to Theorem 3.5. In Phase 3, the scheme in Section 2.2.3 is utilized again to transmit one bit T_2 with helper producing a one-bit quantization of Z_{n+2}.

The achievability follows by noting that the overall transmission rate tends to $R + R_h$ and the help rate to R_h as n tends to infinity.

Finally, the tightness of the lower bound for the Gaussian channel follows from Theorem 2.4.   □

Chapter 4

Listsize Capacity and Cutoff Rate with a Helper

4.1 Introduction

In this chapter, we continue from Chapter 3 and study the listsize capacity and the cutoff rate with help. We show that on the MMANC, the helper raises the listsize capacity to the same level it raises the cutoff rate, and the latter is increased by the rate of help (until the channel saturates). On general continuous channels, however, it becomes tricky due to the cost constraint. The Gaussian channel with decoder assistance is solved, where a similar result is derived.

The following discussions on the preliminaries for the listsize capacity and the cutoff rate are analogous to Section 3.1.

Consider a general DMC $Q_{Y|X}$. A zero-error list decoder returns all the messages that have positive a posteriori probability given the output sequence y^n, i.e., returns $\mathcal{L}_0(y^n)$. The order-ρ listsize capacity $\mathsf{C}_\ell^{(\rho)}$ for $\rho > 0$ is defined as the supremum of achievable rates under the condition

$$\lim_{n\to\infty} \mathbb{E}\big[|\mathcal{L}_0(Y^n)|^\rho\big] = 1. \tag{4.1}$$

Previous studies on the listsize capacity are also scarce, and the listsize capacity shares many properties with the erasures-only capacity: It is neither continuous in the channel nor purely combinatorial; sufficient conditions for the listsize capacity to equal the cutoff rate is known [24]; single-letter characterization remains unknown, while a multi-letter characterization [21, 24] is given by

$$C_\ell^{(\rho)} = \lim_{k \to \infty} \frac{1}{m} \max_{P^{(m)}} \inf_{\substack{\tilde{W}^{(m)}, \hat{W}^{(m)} : \ \hat{W}^{(m)} \ll Q_{Y|X}^m, \\ P^{(m)} \tilde{W}^{(m)} = P^{(m)} \hat{W}^{(m)}}} I\big(P^{(m)}, \hat{W}^{(m)}\big)$$
$$+ D\big(\tilde{W}^{(m)} \big\| Q_{Y|X}^m \big| P^{(m)}\big); \quad (4.2)$$

and it remains unknown even for some simplest channels.

The order-ρ cutoff rate $\mathsf{R}^{(\rho)}_{\text{cutoff}}$ is obtained analogously to $\mathsf{C}_\ell^{(\rho)}$, but with $\mathcal{L}_0(y^n)$ in (4.1) replaced with the at-least-as-likely list $\mathcal{L}(m, y^n)$ with respect to the transmitted message m, i.e.,

$$\lim_{n \to \infty} \mathbb{E}\big[|\mathcal{L}(M, Y^n)|^\rho\big] = 1. \qquad (4.3)$$

Because $\mathcal{L}(m, y^n) \subseteq \mathcal{L}_0(y^n)$ with probability one,

$$\mathsf{C}_\ell^{(\rho)} \leq \mathsf{R}^{(\rho)}_{\text{cutoff}}. \qquad (4.4)$$

The cutoff rate on a DMC can be expressed as [22, 24]

$$\mathsf{R}^{(\rho)}_{\text{cutoff}} = \max_{P_X} \frac{\mathsf{E}_0(\rho, P_X, Q_{Y|X})}{\rho}, \qquad (4.5)$$

where $\mathsf{E}_0(\rho, P_X, Q_{Y|X})$, or simply written as $\mathsf{E}_0(\rho, P_X)$ if the channel is clear from the context, is Gallager's function

$$\mathsf{E}_0(\rho, P_X, Q_{Y|X}) = -\log \sum_{y \in \mathcal{Y}} \bigg(\sum_{x \in \mathcal{X}} P_X(x) \, Q_{Y|X}(y|x)^{\frac{1}{1+\rho}} \bigg)^{1+\rho}. \qquad (4.6)$$

In particular, on the MMANC, the cutoff rate can be computed as:

Lemma 4.1. *For $\rho > 0$, the cutoff rate on the MMANC is*

$$\mathsf{R}^{(\rho)}_{\text{cutoff}} = \log |\mathcal{A}| - H_{\tilde{\rho}}(Q_Z), \qquad (4.7)$$

where $\tilde{\rho} = \frac{1}{1+\rho}$ and $H_{\tilde{\rho}}(\cdot)$ denotes the Rényi entropy of order $\tilde{\rho}$.

Proof. See Appendix B.1. $\square$

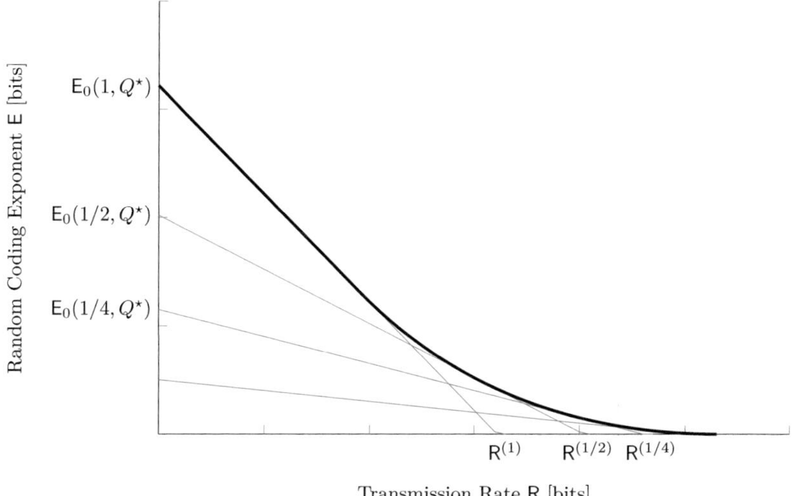

Figure 4.1: Random coding exponent for a BSC. Here $Q^\star = \mathrm{Ber}(1/2)$ is the maximum achieving input distribution for E_0 for any $\rho > 0$.

As a historical remark, perhaps interestingly, the cutoff rate was studied extensively by Arıkan, which inspired his revolutionary invention of the Polar Code [34]–[36]. In fact, there have always been two different perspectives on the cutoff rate (4.5).

The first is related to the error exponent, where the terminology "cutoff rate" often refers simply to the value of $\max_Q \mathsf{E}_0(1, Q)$. It appears as the interception of the curve of the random coding exponent [37, Section 5.6], [38]

$$\mathsf{E}_\mathrm{r}(\mathsf{R}) = \max_{0 \le \rho \le 1} \left\{ -\rho \mathsf{R} + \max_Q \mathsf{E}_0(\rho, Q) \right\} \tag{4.8}$$

with the E-axis, as is illustrated by an example in Fig. 4.1. A generalized operational meaning is introduced by Csiszár [39]. For $0 < \rho \le 1$, he considered the tangent lines of slope $-\rho$ to the Reliability Function under constant composition codes, and defined the cutoff rate $\mathsf{R}^{(\rho)}$ as the maximum interception to the R-axis optimized over the code composition. On a binary symmetric channel (BSC), these relations are depicted in Fig. 4.1. For $\rho > 1$, Csiszár characterized $\mathsf{R}^{(\rho)}$ with the same operational meaning but now with respect to the Reliability Function under list decoding. Under both scenarios, the interception $R^{(\rho)}$ was computed as the RHS of (4.5) [39, Theorem 2].

The second perspective originates from the sequential decoding of tree codes, and the cutoff rate (a.k.a. the *computational* cutoff rate) is defined as the maximal rate under which its computational complexity is bounded [37, Section 6.9], [22]. Equivalently, it can be characterized with the Massey-Arıkan guessing problem [22, 40], where the decoder guesses the transmitted message M based on the output Y^n with a sequence of attempts, and at each time, a genie tells it whether or not the guess is correct. $R^{(\rho)}_{\text{cutoff}}$ is then the maximal rate such that the ρ-th moment of the number of attempts until guessing correctly approaches one. More discussions on the Massey-Arıkan guessing problem can be found in Section 4.2.3.

This second perspective is equivalent to the definition (4.3) (Remark 4.13), and we shall stick to it throughout this chapter. Thus, some previous results on the Reliability Function (e.g., those in [5]) do not carry over directly.

In the presence of feedback, the cutoff rate remains unchanged on a DMC [23], and the listsize capacity is known only if the zero-error capacity C_0 (as is defined in Section 5.1) is positive, as is stated in the following theorem.

Theorem 4.2 ([23, Theorem 1.1]). *For any $\rho > 0$, on every DMC with $C_0 > 0$, the order-ρ listsize capacity equals the order-ρ cutoff rate.*

A proof of Theorem 4.2 that is much simpler than the original one in [23] can be found in Remark 4.7.

4.2   MMANC

Consider the MMANC with a helper. The listsize capacity and the cutoff rate under different types of assistance are defined as the maximal R, for which there exists a sequence of coding schemes of rate-R under rate-R_h help, subject to the conditions (4.1) and (4.3), respectively, where the lists $\mathcal{L}_0$ and $\mathcal{L}$ now correspond to those under each type of assistance, as is analyzed explicitly in Section 2.2.1.

We establish the following theorem, namely, the helper raises the listsize capacity to the same level that it raises the cutoff rate, and the latter is increased by the rate of help.

Theorem 4.3. *For $\rho > 0$, the ρ-th order listsize capacity and the ρ-th order cutoff rate of the MMANC with rate-R_h assistance are given by*

$$\log |\mathcal{A}| - \left\{ H_{\tilde{\rho}}(Q_Z) - R_h \right\}^+, \tag{4.9}$$

irrespective of whether the help is presented to the decoder, to the encoder, or to both.

Remark 4.4. Theorem 4.3 is reminiscent of the analogous result in Theorem 3.1 on the erasures-only capacity, and this is not a coincidence: The Shannon capacity is defined by replacing the zero-error list $\mathcal{L}_0$ in the definition of erasures-only capacity (3.1) with the at-least-as-likely list $\mathcal{L}$. This is analogous to the relation between (4.1) and (4.3). As we shall see, zero-rate assistance already mitigates the difference between $\mathcal{L}_0$ and $\mathcal{L}$. This explains Theorem 3.1 and Theorem 4.3 from a unified point of view. $\triangle$

Theorem 4.3 will be proved via the following three steps in the following subsections. Firstly, the direct part for $C_{\ell,\mathrm{dec}}^{(\rho)}$ is established in Section 4.2.1. Secondly, that for $C_{\ell,\mathrm{enc}}^{(\rho)}$ is proved in Section 4.2.2—unlike the erasures-only capacity, now a more sophisticated coding scheme is required. Lastly, with some preliminaries discussed in Section 4.2.3, the converse part is proved for $R_{\mathrm{cutoff,both}}^{(\rho)}$ in Section 4.2.4,. All the remaining achievability and converse results follow from the facts that $R_{\mathrm{cutoff},\star}^{(\rho)} \geq C_{\ell,\star}^{(\rho)}$ and that assistance never hurts.

In the presence of feedback, we claim that all the capacities remain unchanged.

Theorem 4.5. *On the MMANC in the presence of rate-R_h assistance and a feedback link, the ρ-th order listsize capacity and the ρ-th order cutoff rate are given by*

$$\log |\mathcal{A}| - \left\{ H_{\tilde{\rho}}(Q_Z) - R_h \right\}^+, \tag{4.10}$$

irrespective of whether the help is presented to the decoder, to the encoder, or to both.

Proof. The converse part is essentially the same as Section 4.2.4, with an extra twist following [23, Section II]. The details can be found in Appendix B.2 $\qquad\square$

4.2.1   Direct Part: $\mathsf{C}_{\ell,\mathrm{dec}}^{(\rho)}$

The direct part for decoder assistance follows from a time-sharing argument. Consider the following three cases:

- Case 1: $\mathsf{R}_\mathrm{h} = 0$.

On account of Lemma 4.1, we only need to show that $\mathsf{C}_{\ell,\mathrm{dec}}^{(\rho)}(0) \geq \mathsf{R}_\mathrm{cutoff}^{(\rho)}$, i.e., the cutoff rate in the absence of assistance can be achieved.

To do so, we show how—starting with a sequence of codebooks $\{\mathcal{C}_n\}$ (without help) for which

$$\lim_{n\to\infty} \mathbb{E}\big[|\mathcal{L}(M, Y^n)|^\rho\big] = 1 \tag{4.11}$$

—we can construct a zero-rate helper for which the zero-error list for said codes satisfies

$$\mathbb{E}\big[|\mathcal{L}_0(Y^n, T)|^\rho\big] \leq \mathbb{E}\big[|\mathcal{L}(M, Y^n)|^\rho\big], \tag{4.12}$$

and hence

$$\lim_{n\to\infty} \mathbb{E}\big[|\mathcal{L}_0(Y^n, T)|^\rho\big] = 1. \tag{4.13}$$

We begin by indexing the family $\mathcal{P}_n$ of PMFs on $\mathcal{A}$ with denominator n. To send the message $m \in \mathcal{M}$, the encoder transmits the m-th codeword in $\mathcal{C}_n$, and the helper produces as T the index of the empirical type $\hat{P}_{Z^n}$ of the noise sequence. Since the cardinality of $\mathcal{P}_n$ is subexponential in n, the rate of help is zero.

On our channel (2.11), T determines the conditional probability of Y^n given the message m [25, Theorem 11.1.2]

$$Q_{Y^n|M}(Y^n|m) = Q_{Y|X}^n(Y^n|x^n(m)) \tag{4.14}$$

$$= Q_Z^n(Y^n \ominus x^n(m)) \tag{4.15}$$

$$= 2^{-n(H(\hat{P}_{Z^n}) + D(\hat{P}_{Z^n}\|Q_Z))}, \tag{4.16}$$

so the zero-error list can only contain messages $\tilde{m}$ for which

$$Q_{Y^n|M}(Y^n|\tilde{m}) = Q_{Y^n|M}(Y^n|m), \tag{4.17}$$

and (4.12) follows from the inclusion

$$\mathcal{L}_0(Y^n, T) \subseteq \{\tilde{m} \in \mathcal{M}: Q_{Y^n|M}(Y^n|\tilde{m}) = Q_{Y^n|M}(Y^n|m)\} \tag{4.18}$$

$$\subseteq \{\tilde{m} \in \mathcal{M}: Q_{Y^n|M}(Y^n|\tilde{m}) \geq Q_{Y^n|M}(Y^n|m)\} \tag{4.19}$$

$$= \mathcal{L}(m, Y^n). \tag{4.20}$$

- Case 2: $\mathsf{R_h} > H_{\tilde{\rho}}(Q_Z)$.

The codebook consists of all the sequences in $\mathcal{A}^n$. To send the message $m \in \mathcal{M} = \{1, \ldots, |\mathcal{A}|^n\}$, the encoder transmits the m-th codeword. As to the helper, we rely on a result on Task Encoding [41, Theorem 1.2]: if $\mathsf{R_h} > H_{\tilde{\rho}}(Q_Z)$, then there exists a sequence of mappings

$$\mathsf{h}_n \colon \mathcal{A}^n \to \mathcal{T} = \{1, \ldots, 2^{n\mathsf{R_h}}\} \tag{4.20a}$$

$$z^n \mapsto t, \tag{4.20b}$$

with pre-images $\mathsf{h}_n^{-1}(t) = \{z^n \in \mathcal{A}^n \colon \mathsf{h}_n(z^n) = t\}$ satisfying

$$\lim_{n \to \infty} \mathbb{E}_{Q_Z^n}\left[\left|\mathsf{h}_n^{-1}(\mathsf{h}_n(Z^n))\right|^{\rho}\right] = 1. \tag{4.21}$$

Using this result, let the helper's description of the noise be $T = \mathsf{h}_n(Z^n)$. Based on (Y^n, T), the decoder produces the list $\mathcal{L}_0(Y^n, T)$, and

$$\mathcal{L}_0(Y^n, T) \subseteq \{m \in \mathcal{M} \colon \mathsf{h}_n(Y^n \ominus x^n(m)) = T\}. \tag{4.22}$$

Its ρ-th moment thus satisfies

$$\mathbb{E}\left[|\mathcal{L}_0(Y^n, T)|^{\rho}\right] \leq \mathbb{E}\left[\left|\{m \in \mathcal{M} \colon \mathsf{h}_n(Y^n \ominus x^n(m)) = T\}\right|^{\rho}\right] \tag{4.23}$$

$$\leq \mathbb{E}\left[\left|\{z^n \in \mathcal{A}^n \colon \mathsf{h}_n(z^n) = T\}\right|^{\rho}\right] \tag{4.24}$$

$$- \mathbb{E}\left[\left|\mathsf{h}_n^{-1}(\mathsf{h}_n(Z^n))\right|^{\rho}\right], \tag{4.25}$$

which, by (4.21), tends to 1 as n tends to infinity.

- Case 3: $0 < \mathsf{R_h} \leq H_{\tilde{\rho}}(Q_Z)$.

For any $\delta > 0$, we divide the transmission block into two parts of relative length $\frac{\mathsf{R_h}}{(1+\delta)H_{\tilde{\rho}}(Q_Z)}$ and $1 - \frac{\mathsf{R_h}}{(1+\delta)H(Q_Z)}$ and apply the aforementioned coding schemes for helper rates of $(1+\delta)H_{\tilde{\rho}}(Q_Z)$ and zero, respectively. The overall zero-error list $\mathcal{L}_0$ is the Cartesian product of corresponding lists for the two phases $\mathcal{L}_{0,1}$ and $\mathcal{L}_{0,2}$, and because the channel is memoryless,

$$\lim_{n \to \infty} \mathbb{E}\left[|\mathcal{L}_0|^{\rho}\right] = \lim_{n \to \infty} \mathbb{E}\left[|\mathcal{L}_{0,1} \times \mathcal{L}_{0,2}|^{\rho}\right] \tag{4.26}$$

$$= \lim_{n \to \infty} \mathbb{E}\left[|\mathcal{L}_{0,1}|^{\rho}\right]\mathbb{E}\left[|\mathcal{L}_{0,2}|^{\rho}\right] \tag{4.27}$$

$$= 1. \tag{4.28}$$

Thus, the total rate achieved by this time-sharing scheme is

$$\frac{\mathsf{R_h}\log|\mathcal{A}|}{(1+\delta)H_{\tilde{\rho}}(Q_Z)} + \left(1 - \frac{\mathsf{R_h}}{(1+\delta)H_{\tilde{\rho}}(Q_Z)}\right)\left(\log|\mathcal{A}| - H_{\tilde{\rho}}(Q_Z)\right)$$

$$= \log|\mathcal{A}| - H_{\tilde{\rho}}(Q_Z) + \frac{\mathsf{R_h}}{1+\delta}. \tag{4.29}$$

The result follows by taking $\delta \downarrow 0$. This establishes the direct part for decoder assistance. $\qquad\square$

Remark 4.6. The technique we used to prove achievability in Case 1 can be used to establish that

$$\mathsf{C}_{\text{e-o,dec}} = \mathsf{C}_{\text{dec}} \quad \text{and} \quad \mathsf{C}^{(\rho)}_{\ell,\text{dec}} = \mathsf{R}^{(\rho)}_{\text{cutoff,dec}} \qquad (4.30)$$

on more general DMCs with IID states $\{S_k\}$, provided that Y_k is a function of (x_k, S_k), and S_k is a function of (x_k, Y_k). Indeed, in such channels, the type $\hat{P}_{S^n}$ reveals

$$Q_{Y^n|M}(Y^n|m) = Q^n_{Y|X}(Y^n|X^n(m)) \qquad (4.31)$$
$$= Q^n_S(S^n) \qquad (4.32)$$
$$= 2^{-n(H(\hat{P}_{S^n})+D(\hat{P}_{S^n}\|Q_S))}, \qquad (4.33)$$

Thus, the likelihood function $Q_{Y^n|M}(Y^n|m)$ can be conveyed with negligible overhead. $\qquad\triangle$

Remark 4.7. Using the same technique, we present here a simple proof of Theorem 4.2, that the listsize capacity equals the cutoff rate in the presence of feedback if $\mathsf{C}_0 > 0$.

Proof of Theorem 4.2. Let $\{\mathcal{C}_n\}$ be a sequence of codebooks that achieves the cutoff rate, whose the at-least-as-likely list $\mathcal{L}_n$ satisfies

$$\lim_{n\to\infty} \mathbb{E}\big[|\mathcal{L}_n|^\rho\big] = 1. \qquad (4.34)$$

We propose the following coding scheme. To transmit m, in Phase 1 of length $n_1 = n$, the m-th codeword in $\mathcal{C}_n$ is transmitted. Thanks to the feedback, the output Y^n and hence the joint type $P \triangleq \hat{P}_{(X^n, Y^n)}$ is known to the encoder before Phase 2. In Phase 2, the encoder adopts a zero-error code (of rate at least 1 bit per channel use[1]) to convey the joint type P. The decoder, knowing the joint type, recovers the likelihood function error-free, and the zero-error list $\mathcal{L}_0$ is now a subset of $\mathcal{L}_n$ with probability one. Hence, (4.34) guarantees that

$$\lim_{n\to\infty} \mathbb{E}\big[|\mathcal{L}_0|^\rho\big] \leq \lim_{n\to\infty} \mathbb{E}\big[|\mathcal{L}_n|^\rho\big] = 1, \qquad (4.35)$$

and this implies that $\mathsf{C}^{(\rho)}_{\ell,\text{F}} \geq \mathsf{R}^{(\rho)}_{\text{cutoff,F}}$. $\qquad\square$

$\qquad\qquad\qquad\qquad\qquad\qquad\qquad\qquad\qquad\qquad\qquad\qquad\qquad\triangle$

---

[1]See Section 5.1.

4.2.2 Direct Part: $\mathsf{C}^{(\rho)}_{\ell,\mathrm{enc}}$

We next prove the achievability with encoder assistance. Similarly, we consider three cases:

- Case 1: $\mathsf{R}_\mathrm{h} = 0$.

With negligible rate loss, $\lceil \log |\mathcal{P}_n(\mathcal{A})| / \log |\mathcal{A}| \rceil$ extra channel uses can be used to convey the empirical type of Z^n to the decoder, and the problem reduces to that of decoder assistance as in Section 4.2.1.

- Case 2: $\mathsf{R}_\mathrm{h} > H_{\tilde{\rho}}(Q_Z)$.

Coding is in three phases, with Phase 1 of length $n_1 = n$ using codebook $\mathcal{A}^n$, and Phases 2 and 3 of lengths n_2 and n_3 to be specified later, but both negligible compared to n. The helper can thus describe the noise corrupting the channel in Phases 2 and 3 with negligible overhead, allowing the encoder to subtract the noise and thereby enabling error-free transmission in those two phases.

Observing the Phase 1 noise Z^n, the helper conveys its empirical type $T_2 = \hat{P}_{Z^n}$ to the encoder, who conveys it to the decoder in Phase 2. (Since the number of types is subexponential, both the length of the helper's description T_2 and the number of channel uses n_2 the encoder needs to convey T_2 to the decoder are negligible.) Thereafter, $\hat{P}_{Z^n}$ is known to all parties. Consider two subcases:

(i) $\mathsf{R}_\mathrm{h} \geq H(\hat{P}_{Z^n})$.

In this subcase, the encoder, with the helper's assistance, can subtract the Phase 1 noise corrupting the codeword: Since

$$|\mathcal{T}^{(n)}_{\hat{P}_{Z^n}}| \leq 2^{nH(\hat{P}_{Z^n})} \leq 2^{n\mathsf{R}_\mathrm{h}}, \tag{4.36}$$

the helper can set T_1 to be the index of Z^n in $\mathcal{T}^{(n)}_{\hat{P}_{Z^n}}$, so that T_1 and T_2 jointly determine Z^n, allowing the encoder to subtract Z^n from the codeword in Phase 1. The decoder recovers the codeword as Y_1^n losslessly. Phase 3 is unnecessary and zero padding is applied to exhaust the frame.

(ii) $\mathsf{R}_\mathrm{h} < H(\hat{P}_{Z^n})$.

Fix any $\epsilon, \delta > 0$ independent of all other parameters. In this subcase, the noise samples in only nq locations are described and subtracted, where $q \triangleq \lceil n\mathsf{R}_\mathrm{h}/H(\hat{P}_{Z^n}) \rceil / n$. The issue is how to choose the locations and how to convey them to the decoder with negligible overhead. To this end, we use as location indicator a binary codeword U^n in a small codebook $\mathcal{C}_U(\hat{P}_{Z^n}) \subset \{0,1\}^n$, which is specifically designed for $\hat{P}_{Z^n}$, and each of whose codewords has nq components equal to 1. Prior to transmission, codebooks are designed and agreed upon by all parties for each possible $\hat{P}_{Z^n}$, i.e., for each PMF in $\mathcal{P}_n(\mathcal{A})$. The index of the codeword $U^n \in \mathcal{C}_U(\hat{P}_{Z^n})$ (with the dependence of $\mathcal{C}_U$ on $\hat{P}_{Z^n}$ henceforth made implicit) is described with small overhead to the encoder who conveys it to the decoder in Phase 3. With U^n known to all parties, in Phase 1, the helper can describe the noise samples at the corresponding locations, the encoder subtracts them from the codeword, and the decoder can then be certain of the value of the codeword at these locations. Paramount is that the noise samples at the locations indicated by the codeword have an empirical type that allows their description. This is where the codebook construction is critical.

To explain the coding scheme in detail, we need some notation. Given a binary n-tuple u^n, we define $\mathcal{I}(u^n) \triangleq \{i \in [n] : u_i = 1\}$ and $\bar{u}^n \triangleq 1^n - u^n$, where 1^n denotes the all-one sequence, so $\mathcal{I}(u^n) \sqcup \mathcal{I}(\bar{u}^n) = [n]$, where "$\sqcup$" denotes the disjoint union. If $\mathcal{I} \subset [n]$ is of elements $i_1 < i_2 < \cdots < i_{|\mathcal{I}|}$ and $z^n \in \mathcal{A}^n$, then $z^n(\mathcal{I})$ is the tuple $(z_{i_1}, z_{i_2}, \ldots, z_{i_{|\mathcal{I}|}}) \in \mathcal{A}^{|\mathcal{I}|}$ "picked by $\mathcal{I}$", and in particular, its empirical type is $\hat{P}_{z^n(\mathcal{I})} \in \mathcal{P}_{|\mathcal{I}|}(\mathcal{A})$. For simplicity of notation, $z^n(\mathcal{I}(u^n))$ is also denoted $z^n(u^n)$.

To construct $\mathcal{C}_U$, we pick a joint type $P_{Z,U} \in \mathcal{P}_n(\mathcal{A} \times \{0,1\})$, whose Z-marginal is $\hat{P}_{Z^n}$, whose U-marginal is $\mathrm{Ber}(q)$, and $P_{Z,U} \approx \hat{P}_{Z^n} \circ \mathrm{Ber}(q)$— i.e., under which Z and U are "nearly" independent—in the sense that $P_{Z|U=1}$ is approximately $\hat{P}_{Z^n}$, satisfying

$$\left| H(P_{Z|U=1}) - H(\hat{P}_{Z^n}) \right| < \delta. \tag{4.37}$$

The codebook $\mathcal{C}_U \subset \{0,1\}^n$ we use, as a type-covering [26, Lemma 2.34] rate-distortion codebook for some joint type $P_{Z,U}$, is such that: (1) $|\mathcal{C}_U| = 2^{n\epsilon}$, and (2) to each z^n of type $\hat{P}_{Z^n}$, there corresponds some codeword $u^n \in \mathcal{C}_U$ such that (z^n, u^n) is of type $P_{Z,U}$. The existence of such type $P_{Z,U}$ and codebook $\mathcal{C}_U$ for sufficiently large blocklength n can be found in Appendix B.3,

The coding scheme is the following. The helper sets T_3 to be the index of the codeword $U^n \in \mathcal{C}_U$ whose joint type with the noise Z^n is $P_{Z,U}$. The index of the codeword U^n is described using $n\epsilon$ bits to the encoder who conveys it to the decoder in Phase 3. Since the noise sequence $Z^n(U^n)$ picked by $\mathcal{I}(U^n)$ belongs to $\mathcal{T}_{P_{Z|U=1}}^{(nq)}$, and since, by (4.37),

$$\left| \mathcal{T}_{P_{Z|U=1}}^{(nq)} \right| \leq 2^{nqH(P_{Z|U=1})} \tag{4.38}$$

$$< 2^{nq(H(\hat{P}_{Z^n})+\delta)} \tag{4.39}$$

$$< 2^{(n\mathsf{R}_\mathrm{h}/H(\hat{P}_{Z^n})+1)(H(\hat{P}_{Z^n})+\delta)} \tag{4.40}$$

$$\leq 2^{n(\mathsf{R}_\mathrm{h}+\delta)} \tag{4.41}$$

for sufficiently large n, the helper can convey $Z^n(U^n)$ to the encoder using some universal code. In Phase 1, the encoder subtracts the noise at locations $\mathcal{I}(U^n)$ from the codeword. The decoder, with the knowledge of U^n, recovers these nq components of the codeword that are received losslessly, and, with $P_{Z,U}$, knows that the type of the noise subsequence $Z^n(\bar{U}^n)$ at the remaining $n(1-q)$ components is $P_{Z|U=0}$. Its zero-error list is thus of size

$$|\mathcal{L}_0| = \left|\mathcal{T}_{P_{Z|U=0}}^{(n(1-q))}\right| \leq 2^{n(1-q)H(P_{Z|U=0})} \tag{4.42}$$

$$\leq 2^{nH(\hat{P}_{Z^n})-nqH(P_{Z|U=1})} \tag{4.43}$$

$$< 2^{n(H(\hat{P}_{Z^n})-\mathsf{R}_\mathrm{h}+\delta)}, \tag{4.44}$$

where (4.43) holds by the concavity of entropy and the relation $\hat{P}_{Z^n} = qP_{Z|U=1} + (1-q)P_{Z|U=0}$; and (4.44) follows from (4.37).

The duration of Phases 2 and 3 is

$$n_2 = \left\lceil \frac{\log|\mathcal{P}_n(\mathcal{A})|}{\log|\mathcal{A}|} \right\rceil, \quad n_3 = \left\lceil \frac{n\epsilon}{\log|\mathcal{A}|} \right\rceil. \tag{4.45}$$

to allow the description of $\hat{P}_{Z^n}$ in Phase 2 and of the index of the locator codeword in Phase 3. Thus, the total number of bits of assistance is at most

$$n(\mathsf{R}_\mathrm{h} + \delta) + 2\log|\mathcal{P}_n(\mathcal{A})| + 2n\epsilon + (2\log|\mathcal{A}| + 3), \tag{4.46}$$

where the extra terms in the parenthesis are caused by rounding. When $n \to \infty$, the overall assistance rate and transmission rate tend to $(\mathsf{R}_\mathrm{h} + \delta + 2\epsilon)$ and $\log|\mathcal{A}|$, respectively.

We next prove that the ρ-th moment of the output list size tends to 1. Conditional on $\hat{P}_{Z^n}$—no matter which subcase it corresponds to—we can upper bound the ρ-th moment as

$$\mathbb{E}\left[|\mathcal{L}_0(Y^{n+n_2+n_3})|^\rho \mid \hat{P}_{Z^n}\right] \leq 1 + 2^{n\rho(H(\hat{P}_{Z^n})-\mathsf{R}_\mathrm{h}+\delta)}. \tag{4.47}$$

Hence, taking the expectation of (4.47) over $\hat{P}_{Z^n}$ yields

$$
\begin{aligned}
\mathbb{E}\big[|\mathcal{L}_0(&Y^{n+n_2+n_3})|^\rho\big] \\
&= \mathbb{E}\big[\mathbb{E}\big[|\mathcal{L}_0(Y^{n+n_2+n_3})|^\rho \mid \hat{P}_{Z^n}\big]\big] && (4.48) \\
&\leq \sum_{P_Z \in \mathcal{P}_n(\mathcal{A})} \Big(1 + 2^{n\rho(H(P_Z)-\mathsf{R_h}+\delta)}\Big) \cdot \mathbb{P}\big[\hat{P}_{Z^n} = P_Z\big] && (4.49) \\
&\leq 1 + \sum_{P_Z \in \mathcal{P}_n(\mathcal{A})} 2^{n\rho(H(P_Z)-\mathsf{R_h}+\delta)} \cdot 2^{-nD(P_Z\|Q_Z)} && (4.50) \\
&\leq 1 + |\mathcal{P}_n(\mathcal{A})| \max_{P_Z \in \mathcal{P}_n(\mathcal{A})} 2^{n\rho(H(P_Z)-\rho^{-1}D(P_Z\|Q_Z)-\mathsf{R_h}+\delta)} && (4.51) \\
&\leq 1 + \max_{P_Z \in \mathcal{P}_n(\mathcal{A})} 2^{n\rho(H(P_Z)-\rho^{-1}D(P_Z\|Q_Z)-\mathsf{R_h}+2\delta)} && (4.52) \\
&\leq 1 + \max_{P_Z \in \mathcal{P}(\mathcal{A})} 2^{n\rho(H(P_Z)-\rho^{-1}D(P_Z\|Q_Z)-\mathsf{R_h}+2\delta)}, && (4.53)
\end{aligned}
$$

where in (4.50), we upper bound the probability of each type class [25, Theorem 11.1.4]; (4.52) holds for n sufficiently large because the number of types is subexponential in n. Continuing from (4.53) and using the variational characterization of Rényi entropy that

$$
H_{\frac{1}{1+\rho}}(Q) = \max_{P \in \mathcal{P}(\mathcal{A})} \{H(P) - \rho^{-1}D(P\|Q)\}, \qquad \rho > 0, \quad (4.54)
$$

we conclude that the RHS of (4.53) tends to 1 as n tends to infinity as long as δ is sufficiently small. Finally, letting $\epsilon, \delta \downarrow 0$ finishes the proof.

- Case 3: $0 < \mathsf{R_h} \leq H_{\tilde{\rho}}(Q_Z)$.

  Follows by time sharing. $\qquad\qquad\qquad\qquad\qquad\qquad\qquad\qquad\square$

4.2.3   Intermezzo: Massey-Arıkan Guessing Problem

In this section, we slightly detour from our proof and introduce the Massey-Arıkan guessing problem, or simply the guessing problem. We recap some of its properties that will become useful for the converse proof. For more details, we refer the readers to [22] and [26, Chapter 9].

Let $\mathcal{X}$ be a finite set, and X a chance variable over $\mathcal{X}$ of distribution $Q \in \mathcal{P}(\mathcal{X})$. A guesser aims to find out the realization of X by asking a sequence of questions

$$\text{"Is } X = x_1\text{?"}$$

$$\text{"Is } X = x_2\text{?"}$$

$$\vdots$$

In response, a faithful answer "Yes" or "No" is provided to each guess. Let $\mathsf{G}(x)$ be the number of attempts until $x \in \mathcal{X}$ is guessed (for the first time); the goal is to minimize the expected number of attempts until guessing correctly—or more generally, $\mathbb{E}[\mathsf{G}(X)^\rho]$ for given $\rho > 0$. Thus, without loss of optimality, we avoid repeated guesses and formally define a guessing function as a bijective function

$$\mathsf{G} \colon \mathcal{X} \to [\|\mathcal{X}\|], \tag{4.55}$$

and we say $\mathsf{G}(\cdot)$ is optimal if—among all guessing functions—it minimizes $\mathbb{E}[\mathsf{G}(X)^\rho]$.

The following proposition characterizes the optimal guessing strategy.

Proposition 4.8 ([26, Theorem 9.4]). *A guessing function $\mathsf{G}(\cdot)$ is optimal if, and only if,*

$$\Big(Q(x) > Q(x')\Big) \iff \Big(\mathsf{G}(x) < \mathsf{G}(x')\Big), \qquad x, x' \in \mathcal{X}. \tag{4.56}$$

Consider guessing in the presence of side information: Let $\mathcal{Y}$ also be a finite set, and (X, Y) be generated according to $Q_{X,Y} \in \mathcal{P}(\mathcal{X} \times \mathcal{Y})$. To guess X based on Y, the guesser needs to propose a guessing function $\mathsf{G}(\cdot|\cdot)$ so that for each y, $\mathsf{G}(\cdot|y) \colon \mathcal{X} \to [\|\mathcal{X}\|]$ is a bijective function. It seeks to minimize

$$\mathbb{E}[\mathsf{G}(X|Y)^\rho] = \sum_{y \in \mathcal{Y}} Q_Y(y)\mathbb{E}[\mathsf{G}(X|y)^\rho \,|\, Y = y]. \tag{4.57}$$

and the RHS enables us to characterize the optimal guessing strategy as follows.

Proposition 4.9 ([26, Theorem 9.7]). *It is optimal to guess X based on Y if, after observing $Y = y$, guess according to decreasing order of a posteriori probability $Q_{X|Y}(x|y)$.*

Let $\mathsf{G}^*(\cdot|\cdot)$ and $\mathsf{G}^*(\cdot)$ be optimal guessing functions for guessing X with and without side information Y, respectively. Clearly, $\mathbb{E}[\mathsf{G}^*(X|Y)^\rho] \leq \mathbb{E}[\mathsf{G}^*(X)^\rho]$. On the other hand, the benefit from the side information Y is limited by its support.

Proposition 4.10 ([26, Proposition 9.9]).

$$\mathbb{E}[\mathsf{G}^*(X)^\rho] \leq |\mathcal{Y}|^\rho \, \mathbb{E}[\mathsf{G}^*(X|Y)^\rho] \tag{4.58}$$

The ρ-th moment of the optimal guessing function is characterized by Arıkan as follows.

Theorem 4.11 ([22, Theorem 1]). *Let* $\mathsf{G}^*(\cdot|\cdot)$ *be an optimal guessing function for guessing* X *with side information* Y*. Then,*

$$\frac{1}{(1+\ln|\mathcal{X}|)^\rho}2^{\rho H_{\tilde{\rho}}(X|Y)} \le \mathbb{E}[\mathsf{G}^*(X|Y)^\rho] \le 2^{\rho H_{\tilde{\rho}}(X|Y)}. \qquad (4.59)$$

Here, $\tilde{\rho} = 1/(1+\rho)$ *and* $H_\alpha(X|Y)$ *denotes the conditional Rényi entropy of order* α*.*

Theorem 4.11 can be applied to the guessing of a message transmitted over a noisy channel based on the received output sequence. Assume that a uniformly generated message $M \in \{1,\dots,2^{nR}\}$ is mapped to $\mathbf{X} \in \mathcal{X}^n$, which is sent via a channel $W^{(n)}$ (not necessarily memoryless) and generates $\mathbf{Y} \in \mathcal{Y}^n$. By Theorem 4.11,

$$\mathbb{E}[\mathsf{G}^*(M|\mathbf{Y})^\rho]$$
$$\ge \frac{1}{(1+n\mathsf{R})^\rho} \sum_{\mathbf{y}\in\mathcal{Y}^n} \Big(\sum_{m\in\mathcal{M}} \big(2^{-n\mathsf{R}}W^{(n)}(\mathbf{y}|\mathbf{x}(m))\big)^{\frac{1}{1+\rho}} \Big)^{1+\rho} \qquad (4.60)$$
$$= \frac{1}{(1+n\mathsf{R})^\rho} 2^{n\rho\mathsf{R}-\mathsf{E}_0(\rho,P^{(n)},W^{(n)})} \qquad (4.61)$$

where (4.60) holds as M is uniformly distributed; in (4.61), we introduced the induced input distribution on $\mathcal{X}^n$

$$P^{(n)} = \mathrm{Unif}\{\mathbf{x}(1),\dots,\mathbf{x}(m)\}. \qquad (4.62)$$

We can further lower-bound

$$\mathbb{E}[\mathsf{G}^*(M|\mathbf{Y})^\rho] \ge \frac{1}{(1+n\mathsf{R})^\rho} 2^{n\rho\mathsf{R}-\max_{P^{(n)}} \mathsf{E}_0(\rho,P^{(n)},W^{(n)})}, \qquad (4.63)$$

from which Arıkan established the converse part to (4.5) [22]. As we shall see, this inequality will be useful for us in a similar way.

Remark 4.12 (Guessing M over Continuous Channels). Although (4.63) was derived with respect to discrete random variables, it can be generalized easily to continuous channels over $\mathbb{R}^n$ of density $w(\mathbf{y}|\mathbf{x})$. Let M be mapped to the channel input $\mathbf{X}$, and $\mathbf{Y}$ be the induced output, whose density function is denoted $f_{\mathbf{Y}}$. Then,

$$\mathbb{E}[G^*(M|\mathbf{Y})^\rho]$$

$$= \mathbb{E}[\mathbb{E}[G^*(M|\mathbf{y})^\rho \,|\, \mathbf{Y} = \mathbf{y}]] \tag{4.64}$$

$$\geq \frac{1}{(1 + \ln|\mathcal{M}|)^\rho} \mathbb{E}\left[\mathbb{E}\left[\left(\sum_{m \in \mathcal{M}} Q_{M|\mathbf{Y}}(m|\mathbf{y})^{\frac{1}{1+\rho}}\right)^{1+\rho} \,\middle|\, \mathbf{Y} = \mathbf{y}\right]\right] \tag{4.65}$$

$$= \frac{1}{(1 + \ln|\mathcal{M}|)^\rho} 2^{n\rho R} \,\mathbb{E}\left[\mathbb{E}\left[\left(\int_{\mathbf{x} \in \mathbb{R}^n} \left(\frac{w(\mathbf{y}|\mathbf{x})}{f_{\mathbf{Y}}(\mathbf{y})}\right)^{\frac{1}{1+\rho}} dP^{(n)}(\mathbf{x})\right)^{1+\rho} \,\middle|\, \mathbf{Y} = \mathbf{y}\right]\right] \tag{4.66}$$

$$= \frac{1}{(1 + \ln|\mathcal{M}|)^\rho} 2^{n\rho R} \int_{\mathbf{y} \in \mathbb{R}^n} \left(\int_{\mathbf{x} \in \mathbb{R}^n} w(\mathbf{y}|\mathbf{x})^{\frac{1}{1+\rho}} \, dP^{(n)}(\mathbf{x})\right)^{1+\rho} d\nu \tag{4.67}$$

$$= \frac{1}{(1 + \ln|\mathcal{M}|)^\rho} 2^{n\rho R - n E_0^{(n)}(\rho, P^{(n)}, w)} \tag{4.68}$$

where ν is the Lebesgue measure on $\mathbb{R}^n$; in (4.66), we introduce the probability measure

$$P^{(n)} - \mathrm{Unif}\{\mathbf{x}(1), \ldots, \mathbf{x}(m)\}, \tag{4.69}$$

and in (4.68), we identify the multi-letter Gallager's function $E_0^{(n)}$ as is defined in (4.98) ahead[2]. $\qquad \triangle$

Remark 4.13 (Equivalence of Guessing Decoder and At-Least-As-Likely List). We end this section by briefly explaining why our definition (4.3) is always equivalent to Arıkan's with guessing decoder. Indeed, as Proposition 4.9 suggests, it is optimal to guess based on decreasing a posteriori probability, and hence $G^\star(m|y^n)$ differs from $|\mathcal{L}(m, y^n)|$ only if there is a tie. The influence is up to a constant factor of $(1 + \rho)$, as is indicated in the following lemma, making their difference negligible in light of Lemma 4.22 ahead.

Lemma 4.14. *Let* $a, b \in \mathbb{Z}^+$ *with* $a < b$. *Then, for* $\rho > 0$,

$$\frac{b^\rho}{1 + \rho} \leq \frac{1}{b - a + 1} \sum_{k=a}^{b} k^\rho \leq b^\rho. \tag{4.70}$$

---

[2]In contrast to the discrete case, $E_0^{(n)}$ will be normalized by n. This is for the simplicity of notation in Section 4.3

Proof. The second inequality is obvious. The proof for the first inequality is essentially same as [26, Proof of Theorem 9.6]: note that

$$\sum_{k=a}^{b} k^\rho \geq a^\rho + \int_a^b \xi^\rho \, d\xi \tag{4.71}$$

$$= a^\rho + \frac{b^{\rho+1} - a^{\rho+1}}{1+\rho} \tag{4.72}$$

$$\geq \frac{b^{\rho+1} - a^\rho(a-1)}{1+\rho} \tag{4.73}$$

$$\geq \frac{b^{\rho+1} - b^\rho(a-1)}{1+\rho}, \tag{4.74}$$

and the claim follows by dividing the inequalities by $(b - a + 1)$. $\qquad\square$

$\triangle$

4.2.4   Converse Part: $\mathsf{R}^{(\rho)}_{\text{cutoff,both}}$

In this section, we shall establish that

$$\mathsf{R}^{(\rho)}_{\text{cutoff,both}}(\mathsf{R}_\mathsf{h}) \leq \log |\mathcal{A}| - \left\{ H_{\tilde{\rho}}(Q_Z) - \mathsf{R}_\mathsf{h} \right\}^+. \tag{4.75}$$

Since the cutoff rate is always upper bounded by $\log |\mathcal{A}|$, we focus on the case $0 \leq \mathsf{R}_\mathsf{h} < H_{\tilde{\rho}}(Q_Z)$. Conditional on $T = t$, the channel reduces to the non-IID modulo-additive noise channel $Q_{Y^n|X^n,T=t}$ with the noise distribution $Q_{Z^n|T=t}$, and the encoding function $m \mapsto x^n(m,t)$ can be represented using the codebook $\mathcal{C}_t = \left\{ x^n(m,t) : m \in \mathcal{M} \right\}$.

Fix any sequence of rate-R_h helper and corresponding family of rate-R blocklength-n codebooks $\{\mathcal{C}_t\}_{t \in \mathcal{T}}$. Given t and y^n, construct the optimal guessing function by listing the messages in decreasing order of the likelihood $Q_{Y^n|M,T}(y^n|m,t) = Q_{Y^n|X^n,T}(y^n|x^n(m,t),t)$ with ties resolved arbitrarily, and denote the ranking of message m in the list by $\mathsf{G}(m|y^n,t)$, so

$$|\mathcal{L}(m,y^n,t)| \geq \mathsf{G}(m|y^n,t) \tag{4.76}$$

(where the inequality may be strict due to the ties).

Using Arıkan's inequality on guessing (4.63), the conditional ρ-th moment of the RHS can be lower-bounded by

$$\mathbb{E}\left[\mathsf{G}(M|Y^n,t)^\rho \,\big|\, T=t\right]$$
$$\geq \frac{1}{(1+n\mathsf{R})^\rho}\, 2^{n\rho\mathsf{R}-\max_{P_{X^n}} \mathsf{E}_0(\rho,P_{X^n},Q_{Y^n|X^n,T=t})}, \qquad (4.77)$$

where E_0 is Gallager's function of channel $Q_{Y^n|X^n,T=t}$, the modulo-additive noise channel over length-n superalphabets. By applying Lemma 4.1 to superalphabets of length n,

$$\max_{P_{X^n}} \mathsf{E}_0(\rho,P_{X^n},Q_{Y^n|X^n,T=t}) = \rho\log|\mathcal{A}|^n - \rho H_{\tilde\rho}(Q_{Z^n|T=t}). \qquad (4.78)$$

It follows from (4.77) and (4.78) by averaging over all $t \in \mathcal{T}$ that

$$\mathbb{E}\left[\mathsf{G}(M|Y^n,T)^\rho\right]$$
$$= \sum_{t\in\mathcal{T}} Q_T(t) \cdot \mathbb{E}\left[\mathsf{G}(M|Y^n,t)^\rho \,\big|\, T=t\right] \qquad (4.79)$$
$$\geq \sum_{t\in\mathcal{T}} Q_T(t) \cdot \frac{1}{(1+n\mathsf{R})^\rho}\, 2^{n\rho\mathsf{R}-n\rho\log|\mathcal{A}|+\rho H_{\tilde\rho}(Q_{Z^n|T=t})} \qquad (4.80)$$
$$= \frac{1}{(1+n\mathsf{R})^\rho}\, 2^{n\rho\mathsf{R}-n\rho\log|\mathcal{A}|} \sum_{t\in\mathcal{T}} 2^{\rho H_{\tilde\rho}(Q_{Z^n|T=t})+\log Q_T(t)} \qquad (4.81)$$
$$= \frac{1}{(1+n\mathsf{R})^\rho}\, 2^{n\rho\mathsf{R}-n\rho\log|\mathcal{A}|} \cdot 2^{\rho H_{\tilde\rho}(Z^n|T)}. \qquad (4.82)$$

The RHS of (4.82) cannot tend to one as n tends to infinity unless

$$\mathsf{R} \leq \liminf_{n\to\infty} \log|\mathcal{A}| - \frac{1}{n}H_{\tilde\rho}(Z^n|T) \qquad (4.83)$$
$$\leq \liminf_{n\to\infty} \log|\mathcal{A}| - \frac{1}{n}\left(H_{\tilde\rho}(Z^n) - \log|\mathcal{T}|\right) \qquad (4.84)$$
$$= \log|\mathcal{A}| - H_{\tilde\rho}(Z) + \mathsf{R}_\mathrm{h}, \qquad (4.85)$$

where (4.84) follows from the chain rule of Rényi entropy [42, Theorem 3].

$\square$

Remark 4.15. Rate-R_h assistance increases the cutoff rate on MMANC by at most R_h notwithstanding, its proof relies heavily on the particular structure of the channel. Indeed, for general channels, $\mathsf{R}^{(\rho)}_{\mathrm{cutoff,enc}}(\mathsf{R}_\mathrm{h}) - \mathsf{R}^{(\rho)}_{\mathrm{cutoff}}$ and $\mathsf{R}^{(\rho)}_{\mathrm{cutoff,both}}(\mathsf{R}_\mathrm{h}) - \mathsf{R}^{(\rho)}_{\mathrm{cutoff}}$ can be arbitrarily large, as can be illustrated by an example like the one in [3, Appendix A].

However, it can be proved that

$$R^{(\rho)}_{\text{cutoff,dec}}(R_\text{h}) \leq R_\text{h} + R^{(\rho)}_{\text{cutoff}} \tag{4.86}$$

always holds for general DMCs with rate-R_h decoder assistance, using Proposition 4.10 in the same way as in Section 4.3.3. $\hfill \triangle$

4.3   Gaussian Channel with Decoder Assistance

In this section, we focus on the Gaussian channel of noise variance N and power-P constrained encoders. The channel is determined by the density function

$$w(y|x) = \frac{1}{\sqrt{2\pi \mathsf{N}}} e^{-\frac{(y-x)^2}{2\mathsf{N}}}, \quad x, y \in \mathbb{R}, \tag{4.87}$$

which we extend to n-tuples in a memoryless fashion:

$$w(\mathbf{y}|\mathbf{x}) = \prod_{k=1}^{n} w(y_k|x_k), \quad \mathbf{x}, \mathbf{y} \in \mathbb{R}^n. \tag{4.88}$$

Throughout this section, we use bold letters to denote n-tuples, e.g., $\mathbf{x}$ in place of x^n. For each $m \in \mathcal{M}$, the m-th codeword is denoted $\mathbf{x}_m$, and it is required that

$$\|\mathbf{x}_m\|^2 \leq n\mathsf{P}. \tag{4.89}$$

We define the signal-to-noise ratio (SNR)

$$\mathsf{A} \triangleq \frac{\mathsf{P}}{\mathsf{N}}. \tag{4.90}$$

Throughout this section, we use logarithms of base e to simplify the expressions.

On said channel without help, the listsize capacity is zero, because no codeword is ruled out by any received sequence, so the zero-error list contains all the messages. As for the cutoff rate, we derive the following result.

Theorem 4.16. *For $\rho > 0$, the order-ρ cutoff rate of the additive Gaussian noise channel is given by*

$$R^{(\rho)}_{\text{cutoff}}(\rho) = R^{(\rho)}_0, \tag{4.91}$$

where

$$R_0^{(\rho)} = \frac{1}{2} \ln \frac{1}{2}\left(1 + \frac{A}{1+\rho} + \sqrt{\left(1 - \frac{A}{1+\rho}\right)^2 + \frac{4A}{(1+\rho)^2}}\right)$$

$$+ \frac{1+\rho}{2\rho}\frac{1}{2}\left(1 + \frac{A}{1+\rho} - \sqrt{\left(1 - \frac{A}{1+\rho}\right)^2 + \frac{4A}{(1+\rho)^2}}\right)$$

$$+ \frac{1}{2\rho} \ln \frac{1}{2}\left(1 - \frac{A}{1+\rho} + \sqrt{\left(1 - \frac{A}{1+\rho}\right)^2 + \frac{4A}{(1+\rho)^2}}\right). \quad (4.92)$$

Indeed, $R_0^{(\rho)}$ as is defined (in nats) by (4.92) is a function that plays a prominent role in the analysis of the Reliability Function of said channel [37, Sec. 7.4], [43]. That analysis does not, however, carry over directly to our setting because it deals with error exponents and not lists.

In the presence of a rate-R_h helper that observes the noise sequence and provides a description to the decoder, the zero-error list can now be explicitly expressed as

$$\mathcal{L}_0(\mathbf{y}, t) = \{m \in \mathcal{M} : h(\mathbf{y} \ominus \mathbf{x}_m) = t\}, \quad (4.93)$$

and the at-least-as-likely list

$$\mathcal{L}(m, \mathbf{y}, t) = \{\tilde{m} \in \mathcal{M} : w(\mathbf{y}|\mathbf{x}_{\tilde{m}}) \geq w(\mathbf{y}|\mathbf{x}_m)$$
$$\text{and } h(\mathbf{y} \ominus \mathbf{x}_{\tilde{m}})) = t\}. \quad (4.94)$$

The listsize capacity and the cutoff rate are defined accordingly in the same way as in Section 4.2.

We establish the following result, namely, the listsize capacity and the cutoff rate are both given by the sum of the assistance rate R_h and the order-ρ cutoff rate without a helper.

Theorem 4.17. *On the Gaussian channel, the listsize capacity and the cutoff rate with rate-R_h decoder assistance are given by*

$$C_{\ell,\text{dec}}^{(\rho)}(R_h) = R_{\text{cutoff,dec}}^{(\rho)}(R_h) = R_{\text{cutoff}}^{(\rho)} + R_h \quad (4.95)$$

where $R_{\text{cutoff}}^{(\rho)}$ is the order-ρ cutoff rate of the channel (without assistance) as given in (4.91)–(4.92).

Remark 4.18. Similarly, it is interesting to note that (4.95) also holds when the help rate R_h is zero: the number of help bits required to increase the listsize capacity from zero to $R_{\text{cutoff}}^{(\rho)}$ is sublinear in the blocklength. In fact, as we shall see, all it takes is a sufficiently fine quantization of the normalized squared Euclidean norm of the noise sequence. $\triangle$

The rest of the section is organized as follows. Section 4.3.1 contains some classical and some new observations regarding Gallager's E_0 function and its modification, which will be critical for our discussion. Section 4.3.2 derives the cutoff rate of the Gaussian channel without help and proves Theorem 4.16. Section 4.3.3 proves the converse part to Theorem 4.17 that the cutoff rate can not exceed (4.95). Section 4.3.4 describes and analyzes a coding scheme that proves the direct part for the listsize capacity.

4.3.1   Preliminaries

Given $\rho > 0$ and any probability measure Q on $\mathbb{R}$, Gallager's E_0 function for our channel is defined as [37]

$$E_0(\rho, Q) = -\ln \int_{y \in \mathbb{R}} \left(\int_{x \in \mathbb{R}} w(y|x)^{\frac{1}{1+\rho}} \, dQ(x) \right)^{1+\rho} d\nu(y), \qquad (4.96)$$

where ν is the Lebesgue measure on $\mathbb{R}$.

The result of maximizing $E_0(\rho, Q)$ over all Q under which $\mathbb{E}[X^2] \leq P$, is denoted $E_0^*(\rho)$:

$$E_0^*(\rho) = \sup_{Q \,:\, \int x^2 \, dQ(x) \leq P} E_0(\rho, Q). \qquad (4.97)$$

The multi-letter extension of E_0 is

$$E_0^{(n)}(\rho, Q^{(n)}) = -\frac{1}{n} \ln \int_{\mathbf{y} \in \mathbb{R}^n} \left(\int_{\mathbf{x} \in \mathbb{R}^n} w(\mathbf{y}|\mathbf{x})^{\frac{1}{1+\rho}} \, dQ^{(n)}(\mathbf{x}) \right)^{1+\rho} d\nu(\mathbf{y}),$$
$$(4.98)$$

where $Q^{(n)}$ is a probability measure on $\mathbb{R}^n$; the integrals are over $\mathbb{R}^n$; the channel $w(\mathbf{y}|\mathbf{x})$ is defined in (4.88); and ν is now the Lebesgue measure on $\mathbb{R}^n$. Similarly,

$$E_0^{(n),*}(\rho) = \sup_{Q \,:\, \int \|\mathbf{x}\|^2 \, dQ(\mathbf{x}) \leq nP} E_0^{(n)}(\rho, Q). \qquad (4.99)$$

Given probability measures $Q^{(m)}$ on $\mathbb{R}^m$ and $Q^{(n)}$ on $\mathbb{R}^n$ that satisfy the power constraints $\mathbb{E}\left[\|\mathbf{X}\|^2\right] \leq m\mathsf{P}$ and $\mathbb{E}\left[\|\mathbf{X}\|^2\right] \leq n\mathsf{P}$ respectively, the product measure $Q^{(m)} \times Q^{(n)}$ on $\mathbb{R}^{m+n}$ satisfies the power constraint $\mathbb{E}\left[\|\mathbf{X}\|^2\right] \leq (m+n)\mathsf{P}$ and

$$(m+n)\,\mathsf{E}_0^{(m+n)}\left(\rho, Q^{(m)} \times Q^{(n)}\right) = m\,\mathsf{E}_0^{(m)}\left(\rho, Q^{(m)}\right) + n\,\mathsf{E}_0^{(n)}\left(\rho, Q^{(n)}\right) \tag{4.100}$$

because

$$(m+n)\,\mathsf{E}_0^{(m+n)}\left(\rho, Q^{(m)} \times Q^{(n)}\right)$$

$$= -\ln \int_{\mathbf{y}\in\mathbb{R}^{m+n}} \left(\int_{\mathbf{x}\in\mathbb{R}^{m+n}} w(\mathbf{y}|\mathbf{x})^{\frac{1}{1+\rho}} \, \mathrm{d}\left(Q^{(m)} \times Q^{(n)}\right)(\mathbf{x})\right)^{1+\rho} \mathrm{d}\nu(\mathbf{y}) \tag{4.101}$$

$$= m\,\mathsf{E}_0^{(m)}(\rho, Q^{(m)}) + n\,\mathsf{E}_0^{(n)}(\rho, Q^{(n)}). \tag{4.102}$$

The sequence $\left\{n\,\mathsf{E}_0^{(n),*}(\rho)\right\}$ is thus superadditive, and Feket's Subadditive lemma implies that $\left\{\mathsf{E}_0^{(n),*}(\rho)\right\}$ converges to its supremum:

$$\lim_{n\to\infty} \mathsf{E}_0^{(n),*}(\rho) = \sup_n \mathsf{E}_0^{(n),*}(\rho). \tag{4.103}$$

We shall later see (c.f. (4.130) ahead) that

$$\frac{1}{\rho} \sup_n \mathsf{E}_0^{(n),*}(\rho) = \mathsf{R}_0^{(\rho)}, \tag{4.104}$$

where $\mathsf{R}_0^{(\rho)}$ is defined in (4.92).

We shall also need Gallager's modified E_0 function. To highlight its relation to the unmodified function, which is quite general, we shall use the generic notations $\mathsf{g}(x)$ for x^2 and, $\mathsf{g}(\mathbf{x}) \triangleq \sum_{k=1}^n \mathsf{g}(x_k)$ for $\|\mathbf{x}\|^2$. We shall also replace P with Γ.

Given some $\rho > 0$, some probability distribution Q on $\mathbb{R}$ under which $\mathbb{E}[\mathsf{g}(X)] \leq \Gamma$, and some $r \geq 0$, the modified Gallager's E_0 function $\mathsf{E}_{0,\mathrm{m}}(\rho, Q, r)$ is defined as

$$\mathsf{E}_{0,\mathrm{m}}(\rho, Q, r) = -\ln \int_{y\in\mathbb{R}} \left(\int_{x\in\mathbb{R}} e^{r(\mathsf{g}(x)-\Gamma)} \, w(y|x)^{\frac{1}{1+\rho}} \, \mathrm{d}Q(x)\right)^{1+\rho} \mathrm{d}\nu(y). \tag{4.105}$$

We shall also be interested in the maximum of $\mathsf{E}_{0,\mathrm{m}}(\rho, Q, r)$ over both Q and r. We distinguish between two cases depending on whether

$\mathbb{E}[\mathsf{g}(X)] \leq \Gamma$ holds strictly or not. In the former case, we only allow r to be zero, whereas in the latter case, it can be any nonnegative number. We thus define

$$\mathsf{E}^*_{0,\mathrm{m}}(\rho, Q) = \begin{cases} \sup_{r \geq 0} \mathsf{E}_{0,\mathrm{m}}(\rho, Q, r), & \text{if } \int \mathsf{g}(x)\, \mathrm{d}Q(x) = \Gamma, \\ \mathsf{E}_0(\rho, Q), & \text{if } \int \mathsf{g}(x)\, \mathrm{d}Q(x) < \Gamma, \end{cases} \tag{4.106}$$

and

$$\mathsf{E}^{**}_{0,\mathrm{m}}(\rho) = \sup_{Q:\ \int \mathsf{g}(x)\, \mathrm{d}Q \leq \Gamma} \mathsf{E}^*_{0,\mathrm{m}}(\rho, Q). \tag{4.107}$$

The next proposition provides a lower bound on $\lim \mathsf{E}^{(n),*}_0(\rho)$.

Proposition 4.19. *Any probability distribution Q on $\mathbb{R}$ under which $\mathsf{g}(X)$ is of finite second moment and of expectation Γ provides the lower bound*

$$\lim_{n \to \infty} \mathsf{E}^{(n),*}_0(\rho) \geq \mathsf{E}^*_{0,\mathrm{m}}(\rho, Q). \tag{4.108}$$

Proof. Let Q be any input distributions Q under which $\mathsf{g}(X)$ is of finite second moment and $\mathbb{E}[\mathsf{g}(X)] = \Gamma$. For each $n \in \mathbb{N}$, let $Q^{(n)}$ be the conditional distribution of the n-fold product distribution $Q^{\times n}$ given the event $\{\mathbf{X} \in \mathcal{A}_n\}$, where

$$\mathcal{A}_n = \big\{\mathbf{x} \in \mathbb{R}^n \colon n\Gamma - \delta < \mathsf{g}(\mathbf{x}) \leq n\Gamma\big\} \tag{4.109}$$

where $\delta > 0$ is some positive constant. Thus, for every Borel measurable subset $\mathcal{B}_n$ of $\mathbb{R}^n$,

$$Q^{(n)}(\mathcal{B}_n) = \frac{1}{\mu} Q^{\times n}(\mathcal{B}_n \cap \mathcal{A}_n), \qquad \mathcal{B}_n \in \mathcal{B}(\mathbb{R}^n) \tag{4.110}$$

with

$$\mu = Q^{\times n}(\mathcal{A}_n). \tag{4.111}$$

For any $r \geq 0$, we can upper-bound the Radon–Nykodim derivative of $Q^{(n)}$ with respect to product distribution $Q^{\times n}$ as follows:

$$\frac{\mathrm{d}Q^{(n)}}{\mathrm{d}Q^{\times n}} = \frac{1}{\mu} \mathbb{1}\{\mathbf{x} \in \mathcal{A}_n\} \tag{4.112}$$

$$\leq \frac{1}{\mu} e^{r(\mathsf{g}(\mathbf{x}) - n\Gamma + \delta)} \tag{4.113}$$

$$= \frac{1}{\mu} e^{r\delta}\, e^{r(\mathsf{g}(\mathbf{x}) - n\Gamma)} \tag{4.114}$$

where $\mathbb{1}\{\text{statement}\}$ equals 1 if the statement is true and else 0. Using this bound on the Radon–Nykodim derivative we obtain:

$$\mathsf{E}_0^{(n)}(\rho, Q^{(n)})$$

$$= -\frac{1}{n}\ln\int_{\mathbf{y}\in\mathbb{R}^n}\left(\int_{\mathbf{x}\in\mathbb{R}^n} w(\mathbf{y}|\mathbf{x})^{\frac{1}{1+\rho}}\,\mathrm{d}Q^{(n)}(\mathbf{x})\right)^{1+\rho}\mathrm{d}\nu(\mathbf{y}) \qquad (4.115)$$

$$\geq -\frac{1}{n}\ln\int_{\mathbf{y}\in\mathbb{R}^n}\left(\int_{\mathbf{x}\in\mathbb{R}^n} w(\mathbf{y}|\mathbf{x})^{\frac{1}{1+\rho}}\cdot\frac{1}{\mu}\,e^{r\delta}e^{r(\mathrm{g}(\mathbf{x})-n\Gamma)}\,\mathrm{d}Q^{\times n}(\mathbf{x})\right)^{1+\rho}\mathrm{d}\nu(\mathbf{y}) \qquad (4.116)$$

$$= -\frac{1+\rho}{n}\ln\frac{e^{r\delta}}{\mu} - \ln\int_{y\in\mathbb{R}}\left(\int_{x\in\mathbb{R}} e^{r(\mathrm{g}(x)-\Gamma)}w(y|x)^{\frac{1}{1+\rho}}\,\mathrm{d}Q(x)\right)^{1+\rho}\mathrm{d}\nu(y) \qquad (4.117)$$

$$= -\frac{1+\rho}{n}\ln\frac{e^{r\delta}}{\mu} + \mathsf{E}_{0,\mathrm{m}}(\rho, Q, r). \qquad (4.118)$$

By the Central Limit Theorem, μ tends to $1/2$ as n tends to infinity, so (4.118) implies that

$$\liminf_{n\to\infty}\frac{1}{n}\mathsf{E}_0^{(n)}(\rho, Q^{(n)}) \geq \mathsf{E}_{0,\mathrm{m}}(\rho, Q, r). \qquad (4.119)$$

Taking the supremum of the RHS over all $r \geq 0$ establishes that

$$\liminf_{n\to\infty}\frac{1}{n}\mathsf{E}_0^{(n)}(\rho, Q^{(n)}) \geq \mathsf{E}_{0,\mathrm{m}}^{*}(\rho, Q), \qquad (4.120)$$

and hence, by (4.99), proves (4.108). $\qquad\qquad\qquad\qquad\qquad\square$

We next turn to upper-bounding $\lim \mathsf{E}_0^{(n),*}(\rho)$.

Proposition 4.20. *If the probability distribution $Q^{(n)}$ on $\mathbb{R}^n$ is such that $\mathbb{E}[\mathrm{g}(\mathbf{X})] \leq n\Gamma$, and if f_{R} is any density on $\mathbb{R}$, then*

$$\mathsf{E}_0^{(n)}(\rho, Q^{(n)})$$

$$\leq \sup_{P:\,\int \mathrm{g}(x)\,\mathrm{d}P \leq \Gamma} -(1+\rho)\int_{x\in\mathbb{R}}\ln\left(\int_{y\in\mathbb{R}} w(y|x)^{\frac{1}{1+\rho}}f_{\mathrm{R}}(y)^{\frac{\rho}{1+\rho}}\,\mathrm{d}y\right)\mathrm{d}P(x),$$

$$\qquad (4.121)$$

and consequently,

$$\lim_{n\to\infty} \mathsf{E}_0^{(n),*}(\rho)$$

$$\leq \sup_{P:\ \int \mathbf{g}(x)\,\mathrm{d}P\leq\Gamma} -(1+\rho)\int_{x\in\mathbb{R}} \ln\left(\int_{y\in\mathbb{R}} w(y|x)^{\frac{1}{1+\rho}} f_\mathrm{R}(y)^{\frac{\rho}{1+\rho}}\,\mathrm{d}y\right)\mathrm{d}P(x).$$

$$(4.122)$$

Proof. The proof is based on [44, Proposition 2], which implies that for every density $f_\mathrm{R}^{(n)}$ on $\mathbb{R}^n$ and any probability measure $Q^{(n)}$ on $\mathbb{R}$,

$$n\,\mathsf{E}_0^{(n)}\big(\rho, Q^{(n)}\big)$$

$$\leq -(1+\rho)\int_{\mathbf{x}\in\mathbb{R}^n} \ln\left(\int_{\mathbf{y}\in\mathbb{R}^n} w(\mathbf{y}|\mathbf{x})^{\frac{1}{1+\rho}} f_\mathrm{R}^{(n)}(\mathbf{y})^{\frac{\rho}{1+\rho}}\,\mathrm{d}\mathbf{y}\right)\mathrm{d}Q^{(n)}(\mathbf{x}).$$

$$(4.123)$$

Applying this inequality to the product density

$$f_\mathrm{R}^{(n)}(\mathbf{y}) = \prod_{i=1}^{n} f_\mathrm{R}(y_i), \qquad (4.124)$$

where f_R is a density on $\mathbb{R}$, and using the product form of the channel (4.88), we obtain that for any density f_R on $\mathbb{R}$

$$\mathsf{E}_0^{(n)}(\rho, Q^{(n)})$$

$$\leq -\frac{1}{n}(1+\rho)\sum_{i=1}^{n}\int_{x_i\in\mathbb{R}} \ln\left(\int_{y\in\mathbb{R}} w(y|x_i)^{\frac{1}{1+\rho}} f_\mathrm{R}(y)^{\frac{\rho}{1+\rho}}\,\mathrm{d}y\right)\mathrm{d}Q_i^{(n)}(x_i)$$

$$(4.125)$$

$$= -(1+\rho)\int_{x\in\mathbb{R}} \ln\left(\int_{y\in\mathbb{R}} w(y|x)^{\frac{1}{1+\rho}} f_\mathrm{R}(y)^{\frac{\rho}{1+\rho}}\,\mathrm{d}y\right)\mathrm{d}\bar{Q}(x), \qquad (4.126)$$

where $Q_i^{(n)}$ is the i-th marginal of $Q^{(n)}$, and $\bar{Q}$ is the probability measure on $\mathbb{R}$ defined by

$$\bar{Q} = \frac{1}{n}\sum_{i=1}^{n} Q_i^{(n)}. \qquad (4.127)$$

Observe that if $\mathbb{E}[\mathbf{g}(\mathbf{X})] \leq n\Gamma$ under $Q^{(n)}$, then $\mathbb{E}[\mathbf{g}(X)] \leq \Gamma$ under $\bar{Q}$. This observation and (4.126) establish (4.121). Since (4.121) holds for all n, (4.122) must also hold. $\square$

4.3.2   The Cutoff Rate of the Gaussian Channel

In this section, we prove Theorem 4.16. Since scaling the output does not change the cutoff rate, we will assume without loss of generality that the noise variance is 1 and the transmit power is A. Thus,

$$w(y|x) = \frac{1}{\sqrt{2\pi}} e^{-\frac{(y-x)^2}{2}}, \quad x, y \in \mathbb{R}, \tag{4.128}$$

and each codeword $\mathbf{x}_m$ satisfies

$$\|\mathbf{x}_m\|^2 \leq n\mathsf{A}. \tag{4.129}$$

4.3.2.1   Computing $\lim \mathsf{E}_0^{(n),*}(\rho)$

Here we shall establish that on the Gaussian channel (4.128)

$$\lim_{n\to\infty} \mathsf{E}_0^{(n),*}(\rho) = \rho\mathsf{R}_0^{(\rho)} = \mathsf{E}_{0,\mathrm{m}}^*(\rho, Q_\mathrm{G}), \tag{4.130}$$

where $\mathsf{R}_0^{(\rho)}$ is defined in (4.92), and Q_G is the zero-mean variance-A Gaussian distribution. To this end, we shall derive matching upper and lower bounds on the limit. We begin with the former.

Upper-bounding $\lim_{n\to\infty} \mathsf{E}_0^{(n),*}(\rho)$. We establish the following lemma.

Lemma 4.21. *On the channel* (4.87),

$$\lim_{n\to\infty} \mathsf{E}_0^{(n),*}(\rho) \leq \rho\mathsf{R}_0^{(\rho)}. \tag{4.131}$$

The proof is based on Proposition 4.20 with the density f_R corresponding to a centered Gaussian of carefully chosen variance. The details are postponed to Appendix B.4.

Lower-bounding $\lim \mathsf{E}_0^{(n),*}(\rho)$. To lower-bound $\lim \mathsf{E}_0^{(n),*}(\rho)$, we shall use Proposition 4.19 with Q chosen as a centered variance-A Gaussian, which we denote Q_G. For this probability distribution Gallager calculated $\mathsf{E}_{0,\mathrm{m}}^*(\rho, Q_\mathrm{G})$ [37, Sec. 7.4]. He showed that for any $\rho > 0$,

$$\mathsf{E}_{0,\mathrm{m}}^*(\rho, Q_\mathrm{G}) = \rho\mathsf{R}_0^{(\rho)}, \tag{4.132}$$

where $\mathsf{R}_0^{(\rho)}$ is defined in (4.92). Using this result and Proposition 4.19 we obtain

$$\lim_{n\to\infty} \mathsf{E}_0^{(n),*}(\rho) \geq \mathsf{E}_{0,\mathrm{m}}^*(\rho, Q_\mathrm{G}) \tag{4.133}$$

$$= \rho\mathsf{R}_0^{(\rho)}. \tag{4.134}$$

4.3.2.2   The mapping $\rho \mapsto \mathsf{R}_0^{(\rho)}$ is monotonically decreasing

For the purpose of proving the achievability of $\mathsf{R}_0^{(\rho)}$, we will need the fact that it is monotonically decreasing in ρ. In view of (4.130), it suffices to show that, for every $n \in \mathbb{N}$, the mapping $\rho \mapsto \rho^{-1}\mathsf{E}_0^{(n),*}(\rho)$ is monotonically decreasing. In view of (4.99), the latter will follow once we establish the monotonicity of $\rho \mapsto \rho^{-1}\mathsf{E}_0^{(n)}(\rho, Q^{(n)})$ for any fixed $Q^{(n)}$. Since $\mathsf{E}_0^{(n)}(\rho, Q^{(n)})$ evaluates to zero at $\rho = 0$, this monotonicity can be established by showing that the mapping $\rho \mapsto \mathsf{E}_0^{(n)}(\rho, Q^{(n)})$ is concave. This is established in [37, Appendix 5.B].[3]

4.3.2.3   Achievability of $\mathsf{R}_0^{(\rho)}$

The achievability of $\mathsf{R}_0^{(\rho)}$ will be proved using a random-coding argument. Let Q_G be the zero-mean variance-A Gaussian distribution, let $\delta > 0$ be a positive constant, and let $Q^{(n)}$ be the distribution on $\mathbb{R}^n$ defined in (4.110) and (4.111). Draw the codewords $\{\mathbf{X}_m\}_{m=1,\ldots,e^{n\mathsf{R}}}$ of a blocklength-n random codebook independently, each according to $Q^{(n)}$, so $\|\mathbf{X}_m\|^2 \leq n\mathsf{A}$ with probability 1 for every $m \in \mathcal{M}$. By symmetry, $\mathbb{E}\big[|\mathcal{L}(m, \mathbf{Y})|^\rho\big]$ (where the expectation is over the random choice of the codebook and on the channel behavior) does not depend on m. Consequently,

$$\mathbb{E}\big[|\mathcal{L}(M, \mathbf{Y})|^\rho\big] = \mathbb{E}\big[|\mathcal{L}(1, \mathbf{Y})|^\rho\big] \tag{4.135}$$

and if we establish that $\mathbb{E}\big[|\mathcal{L}(1, \mathbf{Y})|^\rho\big]$ tends to 1, it will follow by the random-coding argument that there exists a codebook for which the left-hand side (LHS) of (4.135)—with the expectation now over the channel behavior only—tends to 1.

     Defining

$$B_m(\mathbf{x}_1, \mathbf{y}) \triangleq \mathbb{1}\{w(\mathbf{y}|\mathbf{X}_m) \geq w(\mathbf{y}|\mathbf{x}_1)\}, \quad \mathbf{x}_1, \mathbf{y} \in \mathbb{R}^n, \tag{4.136}$$

---

[3]That appendix deals with finite alphabets, but the proof goes through also to our case.

we can express the RHS of (4.135) as

$$\mathbb{E}\left[|\mathcal{L}(1,\mathbf{Y})|^{\rho}\right] = \mathbb{E}\left[\left(1 + \sum_{m\neq 1} B_m(\mathbf{X}_1,\mathbf{Y})\right)^{\rho}\right], \tag{4.137}$$

and we seek to show that

$$\lim_{n\to\infty} \mathbb{E}\left[\left(1 + \sum_{m\neq 1} B_m(\mathbf{X}_1,\mathbf{Y})\right)^{\rho}\right] = 1. \tag{4.138}$$

To this end, we shall need the following lemma.

Lemma 4.22. *Let $\{Z_n\}$ be a sequence of random variables taking values in $\mathbb{N}$, and let $\rho > 0$ be fixed. Then, the following two are then equivalent:*

$$\left(\lim_{n\to\infty} \mathbb{E}\left[(1 + Z_n)^{\rho}\right] = 1\right) \iff \left(\lim_{n\to\infty} \mathbb{E}\left[Z_n^{\rho}\right] = 0\right). \tag{4.139}$$

Proof. See Appendix B.5. □

In light of the above lemma, to establish (4.138) it suffices to show that

$$\lim_{n\to\infty} \mathbb{E}\left[\left(\sum_{m\neq 1} B_m(\mathbf{X}_1,\mathbf{Y})\right)^{\rho}\right] = 0, \tag{4.140}$$

i.e., that

$$\lim_{n\to\infty} \mathbb{E}\left[\mathbb{E}\left[\left(\sum_{m\neq 1} B_m(\mathbf{x}_1,\mathbf{y})\right)^{\rho} \,\middle|\, \mathbf{X}_1 = \mathbf{x}_1, \mathbf{Y} = \mathbf{y}\right]\right] = 0, \tag{4.141}$$

where the outer expectation is over $\mathbf{X}_1$ and $\mathbf{Y}$.

A related expectation—but one where it is the conditional expectation that is raised to the ρ-th power—is studied in the following lemma:

Lemma 4.23. *If $\rho > 0$ and $\mathsf{R} < \mathsf{R}_0^{(\rho)}$, then*

$$\lim_{n\to\infty} \mathbb{E}\left[\mathbb{E}\left[\sum_{m\neq 1} B_m(\mathbf{x}_1,\mathbf{y}) \,\middle|\, \mathbf{X}_1 = \mathbf{x}_1, \mathbf{Y} = \mathbf{y}\right]^{\rho}\right] = 0. \tag{4.142}$$

Proof. See Appendix B.6. □

To establish (4.140) using this lemma, we distinguish between two cases depending on whether $0 < \rho \leq 1$ or $\rho > 1$. In the former case, $x \mapsto x^\rho$ is concave, so Jensen's inequality implies that

$$\mathbb{E}\left[\mathbb{E}\left[\left(\sum_{m \neq 1} B_m(\mathbf{x}_1, \mathbf{y})\right)^\rho \bigg| \mathbf{X}_1 = \mathbf{x}_1, \mathbf{Y} = \mathbf{y}\right]\right]$$
$$\leq \mathbb{E}\left[\mathbb{E}\left[\sum_{m \neq 1} B_m(\mathbf{x}_1, \mathbf{y}) \bigg| \mathbf{X}_1 = \mathbf{x}_1, \mathbf{Y} = \mathbf{y}\right]^\rho\right], \qquad (4.143)$$

which, together with Lemma 4.23, implies (4.140) whenever $\mathsf{R} < \mathsf{R}_0^{(\rho)}$.

Suppose now that $\rho > 1$. Conditional on the transmitted codeword $\mathbf{x}_1$ and the output $\mathbf{y}$, the random variables $\{B_m\}_{m \neq 1}$ are IID Bernoulli, with B_m determined by $\mathbf{X}_m$. We can thus use Rosenthal's technique [45, Lemma 5.10][46] to obtain

$$\mathbb{E}\left[\left(\sum_{m \neq 1} B_m(\mathbf{x}_1, \mathbf{y})\right)^\rho \bigg| \mathbf{X}_1 = \mathbf{x}_1, \mathbf{Y} = \mathbf{y}\right]$$
$$\leq 2^{\rho^2} \max\left\{\mathbb{E}\left[\sum_{m \neq 1} B_m(\mathbf{x}_1, \mathbf{y}) \bigg| \mathbf{X}_1 = \mathbf{x}_1, \mathbf{Y} = \mathbf{y}\right]^\rho,\right.$$
$$\left.\mathbb{E}\left[\sum_{m \neq 1} B_m(\mathbf{x}_1, \mathbf{y}) \bigg| \mathbf{X}_1 = \mathbf{x}_1, \mathbf{Y} = \mathbf{y}\right]\right\} \quad (4.144)$$
$$\leq 2^{\rho^2}\left(\mathbb{E}\left[\sum_{m \neq 1} B_m(\mathbf{x}_1, \mathbf{y}) \bigg| \mathbf{X}_1 = \mathbf{x}_1, \mathbf{Y} = \mathbf{y}\right]^\rho\right.$$
$$\left.+\mathbb{E}\left[\sum_{m \neq 1} B_m(\mathbf{x}_1, \mathbf{y}) \bigg| \mathbf{X}_1 = \mathbf{x}_1, \mathbf{Y} = \mathbf{y}\right]\right) \quad (4.145)$$

Taking the expectation over $\mathbf{X}_1$ and $\mathbf{Y}$ yields

$$\mathbb{E}\left[\mathbb{E}\left[\left(\sum_{m \neq 1} B_m(\mathbf{x}_1, \mathbf{y})\right)^\rho \bigg| \mathbf{X}_1 = \mathbf{x}_1, \mathbf{Y} = \mathbf{y}\right]\right]$$
$$\leq 2^{\rho^2} \mathbb{E}\left[\mathbb{E}\left[\sum_{m \neq 1} B_m(\mathbf{x}_1, \mathbf{y}) \bigg| \mathbf{X}_1 = \mathbf{x}_1, \mathbf{Y} = \mathbf{y}\right]^\rho\right]$$
$$+2^{\rho^2} \mathbb{E}\left[\mathbb{E}\left[\sum_{m \neq 1} B_m(\mathbf{x}_1, \mathbf{y}) \bigg| \mathbf{X}_1 = \mathbf{x}_1, \mathbf{Y} = \mathbf{y}\right]\right] \quad (4.146)$$

The first term on the RHS can be treated using the lemma. The second—but for the 2^{ρ^2} constant—is the one encountered when ρ is 1. Since by

Section 4.3.2.2, $\mathsf{R}_0^{(\rho)} \le \mathsf{R}_0^{(1)}$ (because $\rho > 1$ for the case at hand), it too tends to zero when $\mathsf{R} < \mathsf{R}_0^{(\rho)}$.

4.3.2.4 No rate exceeding $\mathsf{R}_0^{(\rho)}$ is achievable

To show the converse, we need Arıkan's lower bound on guessing in Section 4.2.3.

Fix any sequence of rate-R blocklength-n codebooks $\{\mathcal{C}_n\}$ satisfying the cost constraint. For any $n \in \mathbb{N}$, let

$$Q^{(n)} = \mathrm{Unif}\{\mathcal{C}_n\} \tag{4.147}$$

be the induced probability distribution on $\mathbb{R}^n$. Since the codebook satisfies the cost constraint, $\mathbb{E}\big[\|\mathbf{X}\|^2\big] \le n\mathsf{A}$ under $Q^{(n)}$.

Given $\mathbf{y}$, construct the optimal guessing function by listing the messages $m \in \mathcal{M}$ in decreasing order of likelihood $w(\mathbf{y}|\mathbf{x}_m)$ (resolving ties arbitrarily, e.g., ranking low numerical values of m higher), and let $\mathsf{G}(m|\mathbf{y})$ denote the ranking of the message m in this list. Note that

$$|\mathcal{L}(m, \mathbf{y})| \ge \mathsf{G}(m|\mathbf{y}), \tag{4.148}$$

where the inequality can be strict because there may be messages that are in $\mathcal{L}(m, \mathbf{y})$ because they have the same likelihood as m, and that are yet ranked lower than m by $\mathsf{G}(\cdot|\mathbf{y})$ because of the way ties are resolved. It follows from this inequality that the ρ-th moment of $|\mathcal{L}(M, \mathbf{Y})|$ cannot tend to one unless the ρ-th moment of $\mathsf{G}(M|\mathbf{Y})$ does. Using Arıkan's inequality on guessing (4.68),

$$\mathbb{E}\big[\mathsf{G}(M|\mathbf{Y})^\rho\big] \ge \frac{1}{(1+n\mathsf{R})^\rho} \exp\Big(n\rho\mathsf{R} - n\mathsf{E}_0^{(n)}(\rho, Q^{(n)})\Big), \tag{4.149}$$

where $\mathsf{E}_0^{(n)}(\rho, Q^{(n)})$ is defined by (4.98) with the input distribution $Q^{(n)}$ given by (4.147), so the ρ-th moment of $\mathsf{G}(M|\mathbf{Y})$ can tend to one only if

$$\rho\mathsf{R} \le \liminf_{n \to \infty} \mathsf{E}_0^{(n)}(\rho, Q^{(n)}). \tag{4.150}$$

From this, the converse now follows using (4.99) and (4.130) because

$$\liminf_{n \to \infty} \mathsf{E}_0^{(n)}(\rho, Q^{(n)}) \le \lim_{n \to \infty} \mathsf{E}_0^{(n),*}(\rho) \tag{4.151}$$

$$= \rho\mathsf{R}_0^{(\rho)}. \tag{4.152}$$

$\square$

4.3.3 Converse Part of Theorem 4.17

We now prove the converse by showing that

$$R^{(\rho)}_{\text{cutoff,dec}}(R_h) \leq R^{(\rho)}_0 + R_h, \tag{4.153}$$

for which, again, we need Arıkan's lower bound on guessing [22].

Fix any rate-R blocklength-n codebook $\mathcal{C}_n$ and rate-R_h helper. Let $Q^{(n)}$ be the induced distribution on $\mathbb{R}^n$ as is defined in (4.147). Hence, $Q^{(n)}$ satisfies $\mathbb{E}\big[\|\mathbf{X}\|^2\big] \leq nA$.

Given $(\mathbf{y}, t)$, list the messages $m \in \mathcal{M}$ in decreasing order of the a posteriori probability $Q_{M|\mathbf{Y},T}(m|\mathbf{y}, t)$ (resolving ties in some arbitrary fixed way, e.g., ranking low numerical values of m higher), and let $\tilde{G}(m|\mathbf{y}, t)$ denote the ranking of the message m in this list, so

$$|\mathcal{L}(m, \mathbf{y}, t)| \geq \tilde{G}(m|\mathbf{y}, t). \tag{4.154}$$

where the inequality can be strict because of the way ties are resolved. It follows that the ρ-th moment of $|\mathcal{L}(m, \mathbf{y}, t)|$ cannot tend to one unless the ρ-th moment of $\tilde{G}(m|\mathbf{y}, t)$ does. We now establish a necessary condition for the latter:

Create a second list where the messages $m \in \mathcal{M}$ are listed in decreasing order of their a posteriori probability in the absence of help $Q_{M|\mathbf{Y}}(m|\mathbf{y})$ (or equivalently, according to $w(\mathbf{y}|\mathbf{x}(m)))$, and let $G(m|\mathbf{y})$ denote the ranking in that list of the message m given $\mathbf{y}$. The functions $\tilde{G}(m|\mathbf{y}, t)$ and $G(m|\mathbf{y})$ are thus optimal guessing functions with respect to $Q_{M|\mathbf{Y}}$ and $Q_{M|\mathbf{Y},T}$ in the sense of minimal ρ-th moment [26, Theorem 9.4]. Since

$$Q_{M|\mathbf{Y}}(m|\mathbf{y}) = \sum_{t \in \mathcal{T}} Q_{M|\mathbf{Y},T}(m|\mathbf{y}, t)\, Q_T(t), \tag{4.155}$$

it follows from Proposition 4.10 that

$$|\mathcal{T}|^\rho\, \mathbb{E}\big[\tilde{G}(M|\mathbf{Y}, T)^\rho\big] \geq \mathbb{E}\big[G(M|\mathbf{Y})^\rho\big], \tag{4.156}$$

where recall that, as in (4.149), the RHS can be lower-bounded by

$$\mathbb{E}\big[G(M|\mathbf{Y})^\rho\big] \geq \frac{1}{(1 + nR)^\rho}\, \exp\Big(n\rho R - nE^{(n)}_0(\rho, Q^{(n)})\Big). \tag{4.157}$$

Combining (4.156) and (4.157), the ρ-th moment of $\tilde{G}(M|\mathbf{Y}, T)$ can tend to one only if

$$\rho R \leq \liminf_{n \to \infty} \frac{1}{n} \ln |\mathcal{T}|^{\rho} + \mathsf{E}_0^{(n)}(\rho, Q^{(n)}) \tag{4.158}$$

$$\leq \rho \mathsf{R}_{\mathrm{h}} + \rho \mathsf{R}_0^{(\rho)}. \tag{4.159}$$

$\square$

4.3.4   Direct Part of Theorem 4.17

In this section, we prove the direct part of Theorem 4.17: when the decoder can be provided with a rate-R_{h} description of the noise, the listsize capacity can be at least as large as $\mathsf{R}_0^{(\rho)} + \mathsf{R}_{\mathrm{h}}$.

Our proof treats the cases $\mathsf{R}_{\mathrm{h}} = 0$ and $\mathsf{R}_{\mathrm{h}} > 0$ separately. As in Section 4.3.2, we assume that the channel is normalized to having noise variance 1 and transmit power A.

4.3.4.1   Case 1: $\mathsf{R}_{\mathrm{h}} = 0$

Recall that on the MMANC, the analogous result was achieved by letting the helper provide the decoder with a lossless description of the type of the noise sequence. Since this type fully specifies the likelihood function of the transmitted message, the decoder's zero-error list $\mathcal{L}_0(\mathbf{Y}, T)$ contains only messages whose a posteriori probability is equal to that of the transmitted message. The list is therefore a subset of the at-least-as-likely list $\mathcal{L}(M, \mathbf{Y})$ (without help) and hence of smaller-or-equal ρ-th moment. Consequently, any rate that allows the latter to tend to one also allows the former to tend to one.

On the Gaussian channel the likelihood function $w(\mathbf{y}|\mathbf{x}_m)$ is specified by the normalized squared Euclidean norm of the noise sequence $\|\mathbf{z}\|^2/n$. The latter, however, cannot be described at zero rate with infinite precision. This motivates us to quantize it and have the quantized version be the zero-rate help. The result will then follow by considering the high-resolution limit of the achievable rates. For this purpose, a uniform quantizer will do.

Given some large $\mathsf{M} > 0$ (which determines the overload region) and some large K (corresponding to the number of quantization cells), we partition the interval $[0, \mathsf{M}]$ into K subintervals, each of length $\Delta = \mathsf{M}/\mathsf{K}$.

The helper, upon observing the noise sequence $\mathbf{Z}$, produces

$$T = \mathsf{h}(\mathbf{Z}) = \begin{cases} \left\lceil \frac{\|\mathbf{Z}\|^2}{n\Delta} \right\rceil & \text{if } \|\mathbf{Z}\|^2 < n\mathsf{M}, \\ 0 & \text{otherwise,} \end{cases} \tag{4.160}$$

The constant M, is chosen large enough so that the large-deviation probability of overload $\mathbb{P}[\|\mathbf{Z}\|^2 \geq n\mathsf{M}]$ decays sufficiently fast in n to guarantee that—even if an overload results in the list containing all $e^{n\mathsf{R}}$ codewords—the contribution of the overload to the ρ-th moment of the list be negligible:

$$\lim_{n\to\infty} e^{n\rho\mathsf{R}} \, \mathbb{P}[\|\mathbf{Z}\|^2 \geq n\mathsf{M}] = 0. \tag{4.161}$$

The existence of such a constant follows, for example, from the Chernoff bound on the tail of the χ^2-distribution

$$\mathbb{P}[\|\mathbf{Z}\|^2 \geq nt] \leq \exp\left(-n\left(\frac{1}{2}t - \frac{1}{2} - \frac{1}{2}\ln t\right)\right) \tag{4.162}$$

which shows that, for a given $\mathsf{R} < \mathsf{R}_0^{(\rho)}$, we can choose $\mathsf{M} = \max\{2, 10\rho\mathsf{R}_0^{(\rho)}\}$. Since the help takes values in the finite set $\mathcal{T}_n = \{0, 1, \ldots, \mathsf{K}\}$, where K does not depend on the blocklength, it is of zero rate.

As in Section 4.3.2.3, we consider a random codebook $\{\mathbf{X}_m\}_{m=1,\ldots,e^{n\mathsf{R}}}$ whose codewords are drawn independently from the conditional Gaussian distribution, i.e., from $Q^{(n)}$ as is defined in (4.110) and (4.111) with Q being Q_{G}, the centered variance-A Gaussian distribution. Using the same symmetry arguments, we also assume that the transmitted message is $m = 1$ and study the ρ-th moment of the list under this assumption. Defining the random variable

$$V_m(\mathbf{x}_1, \mathbf{y}) = \mathbb{1}\{\mathsf{h}(\mathbf{y} - \mathbf{X}_m) = \mathsf{h}(\mathbf{y} - \mathbf{x}_1)\}, \tag{4.163}$$

for each $\mathbf{x}_1, \mathbf{y} \in \mathbb{R}^n$, and $m \in \mathcal{M}$, we can express the ρ-th moment of the zero-error list when $m = 1$ as

$$\mathbb{E}\left[|\mathcal{L}_0(\mathbf{Y}, T)|^\rho\right] = \mathbb{E}\left[\left(1 + \sum_{m\neq 1} V_m(\mathbf{X}_1, \mathbf{Y})\right)^\rho\right]. \tag{4.164}$$

In view of Lemma 4.22, we need to prove that

$$\lim_{n\to\infty} \mathbb{E}\left[\left(\sum_{m\neq 1} V_m(\mathbf{X}_1, \mathbf{Y})\right)^\rho\right] = 0, \tag{4.165}$$

where the expectation is over both the random choice of the codebook and the channel behavior.

To analyze the RHS of (4.165), we distinguish between whether $h(\mathbf{y} - \mathbf{x}_1)$ differs from 0 (no overload) or equals 0 (corresponding to quantizer overload): In the former case, we define for every $\mathbf{x}_1, \mathbf{y} \in \mathbb{R}^n$ and every message $m \neq 1$ the binary random variable

$$B_m(\mathbf{x}_1, \mathbf{y}; \Delta) = \mathbb{1}\left\{ w(\mathbf{y}|\mathbf{X}_m) \geq w(\mathbf{y}|\mathbf{x}_1) \cdot e^{-\frac{n\Delta}{2}} \right\}, \qquad (4.166)$$

which is an upper bound for $V_m(\mathbf{x}_1, \mathbf{y})$ because

$$V_m(\mathbf{x}_1, \mathbf{y}) = \mathbb{1}\left\{ h(\mathbf{y} - \mathbf{X}_m) = h(\mathbf{y} - \mathbf{x}_1) \right\} \qquad (4.167)$$
$$\leq \mathbb{1}\left\{ \left| \|\mathbf{y} - \mathbf{X}_m\|^2 - \|\mathbf{y} - \mathbf{x}_1\|^2 \right| \leq n\Delta \right\} \qquad (4.168)$$
$$\leq \mathbb{1}\left\{ \|\mathbf{y} - \mathbf{X}_m\|^2 \leq \|\mathbf{y} - \mathbf{x}_1\|^2 + n\Delta \right\} \qquad (4.169)$$
$$= \mathbb{1}\left\{ e^{-\frac{\|\mathbf{y} - \mathbf{X}_m\|^2}{2}} \geq e^{-\frac{\|\mathbf{y} - \mathbf{x}_1\|^2}{2}} \cdot e^{-\frac{n\Delta}{2}} \right\} \qquad (4.170)$$
$$= B_m(\mathbf{x}_1, \mathbf{y}; \Delta), \qquad (4.171)$$

where (4.168) holds because, for the case at hand, the equality of helper's description implies that $\|\mathbf{y} - \mathbf{X}_m\|^2$ and $\|\mathbf{y} - \mathbf{x}_1\|^2$ lie in a same interval of length $n\Delta$. While in the latter case—which is sufficiently rare—we simply upper bound $V_m(\mathbf{x}_1, \mathbf{y})$ by 1.

The ρ-th moment of the list can now be expressed using the Law of Total Expectation as

$$\mathbb{E}\left[\left(\sum_{m \neq 1} V_m(\mathbf{X}_1, \mathbf{Y}) \right)^{\rho} \right]$$
$$= \mathbb{E}\left[\left(\sum_{m \neq 1} V_m(\mathbf{X}_1, \mathbf{Y}) \right)^{\rho} \middle| T \neq 0 \right] \mathbb{P}[T \neq 0]$$
$$+ \mathbb{E}\left[\left(\sum_{m \neq 1} V_m(\mathbf{X}_1, \mathbf{Y}) \right)^{\rho} \middle| T = 0 \right] \mathbb{P}[T = 0]$$
$$\qquad (4.172)$$
$$\leq \mathbb{E}\left[\left(\sum_{m \neq 1} B_m(\mathbf{X}_1, \mathbf{Y}; \Delta) \right)^{\rho} \middle| T \neq 0 \right] \mathbb{P}[T \neq 0] + e^{n\rho R}\, \mathbb{P}[T = 0]$$
$$\qquad (4.173)$$
$$\leq \mathbb{E}\left[\left(\sum_{m \neq 1} B_m(\mathbf{X}_1, \mathbf{Y}; \Delta) \right)^{\rho} \right] + e^{n\rho R}\, \mathbb{P}[T = 0]. \qquad (4.174)$$

The second term on the RHS of (4.174) tends to zero by (4.161). The first term is studied in the following lemma:

Lemma 4.24. *If $\rho > 0$, $\Delta > 0$, and $\mathsf{R} < \mathsf{R}_0^{(\rho)} - \Delta$, then*

$$\lim_{n \to \infty} \mathbb{E}\left[\left(\sum_{m \neq 1} B_m(\mathbf{X}_1, \mathbf{Y}; \Delta) \right)^{\rho}\right] = 0. \tag{4.175}$$

Proof. See Appendix B.7. $\qquad\qquad\qquad\qquad\qquad\qquad\qquad\qquad\square$

For a given $\mathsf{R} < \mathsf{R}_0^{(\rho)}$, achievability is thus established using this lemma and (4.174) by picking M sufficiently large for (4.161) to hold, and then picking K large enough to guarantee that $\mathsf{R} < \mathsf{R}_0^{(\rho)} - \mathsf{M}/\mathsf{K}$ so that, by Lemma 4.24, the first term on the RHS of (4.174) will also tend to zero.

4.3.4.2   Case 2: $\mathsf{R}_\mathrm{h} > 0$

The key to proving the achievability of $\mathsf{R}_{\mathrm{cutoff}}^{(\rho)}(\rho) + \mathsf{R}_\mathrm{h}$ is in showing that rate-R_h help can be utilized to increase the data rate by R_h. To that end, we propose the following blocklength-$(n+1)$ two-phase time-sharing coding scheme.

In the first channel use, we adopt a coding scheme as described in Section 2.2.3, where the helper produces $T_1 \in \mathcal{T}_1 = \{1, \ldots, \lceil e^{n\mathsf{R}_\mathrm{h}} \rceil\}$, and the input symbols take values only in $\mathcal{I} = [0, \sqrt{\mathsf{A}}]$ to convey a message taking values in $\mathcal{T}_1$ error-free. In the remaining n channel uses, the helper operates at rate zero as in Section 4.3.4.1, and thus for any $\mathsf{R} < \mathsf{R}_0^{(\rho)}$, there exists a sequence of blocklength-n rate-R codebooks $\{\mathbf{x}_m\}_{m=1,\ldots,e^{n\mathsf{R}}}$, with $\|\mathbf{x}_m\|^2 \leq n\mathsf{A}$ for every m, and zero-rate helpers $T_2 = \mathsf{h}(Z_2^{n+1})$, such that the zero-error list $\mathcal{L}(Y_2^{n+1}, T_1)$ satisfies

$$\lim_{n \to \infty} \mathbb{E}\left[\left| \mathcal{L}_0(Y_2^{n+1}, T_2) \right|^{\rho}\right] = 1. \tag{4.176}$$

Since Phase 1 is error-free, the overall zero-error list has the same cardinality as that of the second phase, namely $\left| \mathcal{L}_0(Y_2^{n+1}, T_2) \right|$, and hence, its ρ-th moment tends to 1. $\qquad\qquad\square$

Chapter 5

Zero-Error Capacity on MMANCs with a Helper

5.1 Introduction

This chapter investigates the extent to which the zero-error capacity can benefit from a helper. We focus on MMANCs, where the key role will be played by the support set $\mathcal{S}$ of Q_Z

$$\mathcal{S} = \{z \in \mathcal{A} \colon Q_Z(z) > 0\}. \tag{5.1}$$

Example 5.1. In the context of zero-error capacity, when $|\mathcal{S}| = 2$, the MMANCs corresponding to $|\mathcal{A}|$ being 3, 5, and 7 are called, respectively, the Triangle channel, Shannon's Pentagon channel [47], and the Heptagon channel (a.k.a. the 3/2, 5/2, and 7/2 channels, respectively). See Fig. 5.1. $\Diamond$

In the presence of a noiseless feedback link from the receiver to the encoder, we calculate the zero-error helper capacity (Theorem 5.6). In its absence, we derive upper and lower bounds on the zero-error helper capacity (Theorem 5.8) and establish a positivity result for the zero-error capacity (Corollary 5.10): if the assistance rate is positive, then so is the capacity; otherwise, the capacity is positive if, and only if, the support $\mathcal{S}$ of the noise is a strict subset of $\mathcal{A}$. When the cardinality of $\mathcal{A}$ is prime (as in Example 5.1) we calculate the zero-error helper capacity in

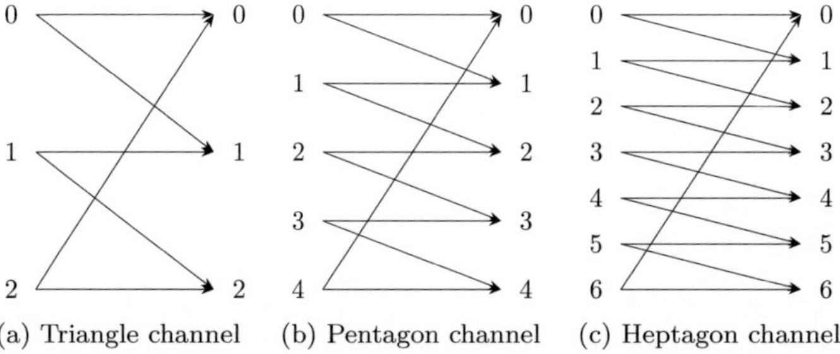

Figure 5.1: Examples of MMANCs

Theorem 5.7 using structured codes. Calculating the zero-error helper capacity without feedback when $|\mathcal{A}|$ is not prime is left as an open problem.

Perhaps paradoxically, the zero-error helper capacity can be calculated as a function of the assistance rate even for some channels whose no-help zero-error capacity is unknown. This is not a contradiction, because a zero-rate description is not tantamount to no description: it still allows for a binary description of length that is sublinear in the blocklength. In fact, the solution of the zero-rate help case is the key to the general solution.

The rest of the chapter is organized as follows. Section 5.2 defines the key quantities of interest and surveys some of the literature that touches on this work. Section 5.3 presents the main results and some of their consequences. The proof of Theorem 5.6 pertaining to feedback is presented in Section 5.4. The proofs of Theorems 5.7 and 5.8 pertaining to the no-feedback setting are presented in Sections 5.5.1 and 5.5.2 respectively.

Throughout this chapter, we use the following notations: If $\mathcal{B} \subseteq \mathcal{A}^n$ is a set of n-tuples, then $\mathcal{B}^\star$ denotes $\mathcal{B} \setminus \{\mathbf{0}\}$, i.e., $\mathcal{B}$ without the all-zero n-tuple. For $\mathcal{B}, \mathcal{B}' \subseteq \mathcal{A}^n$, we denote the sumset and the difference set by

$$\mathcal{B} \oplus \mathcal{B}' = \left\{ \mathbf{b} \oplus \mathbf{b}' : \mathbf{b} \in \mathcal{B},\ \mathbf{b}' \in \mathcal{B}' \right\} \tag{5.2}$$

$$\mathcal{B} \ominus \mathcal{B}' = \left\{ \mathbf{b} \ominus \mathbf{b}' : \mathbf{b} \in \mathcal{B},\ \mathbf{b}' \in \mathcal{B}' \right\}; \tag{5.3}$$

and for $\mathbf{x} \in \mathcal{A}^n$, we write $\mathbf{x} \oplus \mathcal{B}$ and $\mathbf{x} \ominus \mathcal{B}$ for $\{\mathbf{x}\} \oplus \mathcal{B}$ and $\{\mathbf{x}\} \ominus \mathcal{B}$.

5.2 Preliminaries

A blocklength-n code for a (general) DMC $Q_{Y|X}$ with input alphabet $\mathcal{X}$ and output alphabet $\mathcal{Y}$, consists of a message set $\mathcal{M} = \{1, 2, \ldots, |\mathcal{M}|\}$ and an encoding function $f \colon \mathcal{M} \to \mathcal{X}^n$, $m \mapsto \mathbf{x}(m)$. Since the codewords $\mathbf{x}(1), \ldots, \mathbf{x}(|\mathcal{M}|)$ need not be distinct, the codebook $\mathcal{C} = \{\mathbf{x}(1), \mathbf{x}(2), \ldots, \mathbf{x}(|\mathcal{M}|)\}$ is a multiset (i.e., an unordered collection of elements that may repeat) of cardinality $|\mathcal{M}|$.

The zero-error capacity C_0 [47] is the supremum of rates R for which there exists a sequence of rate-R codes, indexed by the blocklength, that fulfill the zero-error requirement that to every output sequence $\mathbf{y} \in \mathcal{Y}^n$ there correspond at most one compatible message, i.e., a message $m \in \mathcal{M}$ for which

$$Q_{Y|X}^n(\mathbf{y}|\mathbf{x}(m)) > 0. \tag{5.4}$$

A necessary and sufficient condition for C_0 to be positive is that there exist channel inputs $x, x' \in \mathcal{X}$ such that $Q_{Y|X}(y|x) \cdot Q_{Y|X}(y|x') = 0$ for every $y \in \mathcal{Y}$ [47]. This characterization can be used, for example, to conclude that C_0 is zero for the Triangle channel (see Example 5.1). It also shows that, whenever C_0 is positive, we can transmit a bit by using the channel once (with the input x or x'). Consequently, C_0 cannot be positive yet smaller than one. As we shall see, this is not the case in the presence of help (Remark 5.13).

Determining the zero-error capacity for general DMCs is an open combinatorial problem and is one of the holy grails of information theory. It is known for some specific channels, including the Pentagon channel: Shannon showed that $\frac{1}{2}\log 5 \le \mathsf{C}_0 \le \log \frac{5}{2}$ in 1959 [47], and Lovász proved, using algebraic graph theory, that the lower bound is tight in 1979 [48]. The zero-error capacity of the Heptagon channel is to date unknown.

The problem is greatly simplified if the time-i channel input may depend not only on the message m but also on the past channel outputs y^{i-1} that are revealed to the encoder via a feedback link from the channel output to the encoder. The zero-error feedback capacity $\mathsf{C}_{0\mathrm{F}}$ is defined like C_0 except that $\mathbf{x}(m)$ in (5.4) is replaced by $\mathbf{x}(m, \mathbf{y}) = (x_1(m), x_2(m, y_1), \ldots, x_n(m, y^{n-1}))$. Since the encoder may ignore the feedback link,

$$\mathsf{C}_{0\mathrm{F}} \ge \mathsf{C}_0. \tag{5.5}$$

The zero-error feedback capacity C_{0F} was determined by Shannon:

Theorem 5.2 ([47]). *On a DMC, if $\mathsf{C}_0 = 0$, then the zero-error feedback capacity C_{0F} is also zero. Else, $\mathsf{C}_{0F} = -\log \pi_0$, where*

$$\pi_0 = \min_{P \in \mathcal{P}(\mathcal{X})} \max_{y \in \mathcal{Y}} \sum_{x \in \mathcal{X}_y} P(x), \tag{5.6}$$

and $\mathcal{X}_y$ comprises the inputs that can induce the output letter y with positive probability:

$$\mathcal{X}_y = \{x \in \mathcal{X} : Q_{Y|X}(y|x) > 0\}. \tag{5.7}$$

Note that, since $\mathsf{C}_{0F} > 0$ if and only if $\mathsf{C}_0 > 0$, and since $\mathsf{C}_{0F} \geq \mathsf{C}_0$, also C_{0F} cannot be positive yet strictly smaller than one. We shall see that this is not true in the presence of zero-rate help (Remark 5.13).

Henceforth, we focus on MMANCs. Applying Theorem 5.2 to the channel yields the following corollary, whose proof can be found in Appendix C.1.

Corollary 5.3. *On the MMANC, if $\mathsf{C}_0 = 0$, then the zero error feedback capacity C_{0F} is also zero. Else,*

$$\mathsf{C}_{0F} = \log |\mathcal{A}| - \log |\mathcal{S}|. \tag{5.8}$$

Now consider a rate-R_h helper. The zero-error helper capacities can be defined as the maximal achievable rate such that to every output $\mathbf{y} \in \mathcal{A}^n$—and also every noise description $t \in \mathcal{T}$, if assistance is presented to the decoder—there exists only one compatible message that could have produced it. To simplify the discussion, we introduce the following assumption.

Assmption 5.4. *We shall assume that*

$$|\mathcal{S}| > 1 \tag{5.9}$$

and

$$\mathsf{R}_h \leq \log |\mathcal{S}|. \tag{5.10}$$

If $|\mathcal{S}| = 1$, then the noise is deterministic and even without feedback or help, the zero-error capacity is $\log |\mathcal{A}|$. And, for any sequence of helpers $h \colon \mathcal{A}^n \to \mathcal{T}$ of rate $\mathsf{R}_h > \log |\mathcal{S}|$, one can construct another sequence of helpers h' of rate $\mathsf{R}'_h = \log |\mathcal{S}|$ with $h'(\mathbf{z}) = h(\mathbf{z})$ for all $\mathbf{z} \in \mathcal{S}^n$, so that $h(\mathbf{Z})$ and $h'(\mathbf{Z})$ are identical with probability one.

5.2.1 Related Work

Related to this work is the following theorem [49] on the case where the noise is of full support:

Theorem 5.5 (MMANC with Noise of Full Support [49]). *On the MMANC with rate-R_h decoder or encoder assistance, if $\mathcal{S} = \mathcal{A}$, then*

$$\mathsf{C}_{0,\mathrm{dec}}(\mathsf{R}_h) = \mathsf{C}_{0,\mathrm{enc}}(\mathsf{R}_h) = \mathsf{R}_h, \tag{5.11a}$$

and in particular,

$$\mathsf{C}_{0,\mathrm{dec}}(0) = 0 \tag{5.11b}$$

and

$$\mathsf{C}_{0,\mathrm{enc}}(0) = 0. \tag{5.11c}$$

This work extends Theorem 5.5 by studying the general case where the noise need not be of full support. As we shall see ahead (Corollary 5.10), the condition $\mathcal{S} = \mathcal{A}$ is also necessary for (5.11b) to hold, and likewise for (5.11c). We also study the effect of feedback on the zero-error helper capacity.

Also related to our results is the work of Merhav on error exponents [5]. To see the relevance, note that on a DMC, the Reliability Function $\mathsf{E}(\mathsf{R})$ equals infinity if, and only if, R can be achieved with zero error. The intuition is the following. Let the transition matrix be $Q_{Y|X}(\cdot|\cdot)$ and the input sequence be $\mathbf{x}$, then all output sequences of positive probability have probability of at least α^n, where

$$\alpha = \min_{(x,y)\in\mathcal{X}\times\mathcal{Y}:\ Q_{Y|X}(y|x)>0} Q_{Y|X}(y|x). \tag{5.12}$$

Consequently,

$$\Big(\mathbb{P}[\mathrm{error}\,|\,M = m] > 0\Big) \implies \Big(\mathbb{P}[\mathrm{error}\,|\,M = m] \geq \alpha^n\Big), \tag{5.13}$$

whose contrapositive can be rewritten as

$$\Big(-\frac{1}{n}\log\mathbb{P}[\mathrm{error}\,|\,M = m] > \log\frac{1}{\alpha}\Big) \implies \Big(\mathbb{P}[\mathrm{error}\,|\,M = m] = 0\Big). \tag{5.14}$$

For our MMANC, Merhav [5, Eq. (57)] derived an upper bound on the Reliability Function with encoder assistance for $\mathsf{R} < \log|\mathcal{A}|$,

$$\mathsf{E}(\mathsf{R}) \leq \min_{\tilde{Q}_Z:\ H(\tilde{Q}_Z)>\log|\mathcal{A}|+\mathsf{R}_h-\mathsf{R}} D(\tilde{Q}_Z\|Q_Z). \tag{5.15}$$

This upper bound is finite if, and only if, $R > \log|\mathcal{A}| - \log|\mathcal{S}| + R_h$: only in this range of rates, there exists a PMF $\tilde{Q}_Z$ that is feasible in the minimization and satisfies $\mathrm{supp}(\tilde{Q}_Z) \subseteq \mathcal{S}$. This implies an upper bound on the zero-error capacity

$$C_{0,\mathrm{enc}}(R_h) \leq \log|\mathcal{A}| - \log|\mathcal{S}| + R_h. \tag{5.16}$$

When $|\mathcal{A}|$ is a prime, our results (Theorem 5.7) show that this upper bound is tight.

5.3   Main Results

In the *presence* of feedback, as in Fig. 1.4, following theorem addresses the zero-error helper capacity of the MMANC in the presence of a feedback link.

Theorem 5.6 (Assistance and Feedback). *On the MMANC with feedback and rate-R_h assistance,*

$$C_{0F,\mathrm{both}}(R_h) = C_{0F,\mathrm{enc}}(R_h) = C_{0F,\mathrm{dec}}(R_h) \tag{5.17}$$
$$= \log|\mathcal{A}| - \log|\mathcal{S}| + R_h. \tag{5.18}$$

In the *absence* of feedback, as in Fig. 1.2, we distinguish two different cases. When the cardinality $|\mathcal{A}|$ of the alphabet $\mathcal{A}$ is prime, the zero-error helper capacity without feedback is determined in the following theorem:

Theorem 5.7 (Assistance without Feedback: Prime Cardinality). *On the MMANC with rate-R_h assistance, if $|\mathcal{A}|$ is prime, then*

$$C_{0,\mathrm{both}}(R_h) = C_{0,\mathrm{enc}}(R_h) = C_{0,\mathrm{dec}}(R_h) \tag{5.19}$$
$$= \log|\mathcal{A}| - \log|\mathcal{S}| + R_h. \tag{5.20}$$

When $|\mathcal{A}|$ is not necessarily a prime, we provide the following upper and lower bounds:

Theorem 5.8 (Assistance without Feedback: General Cardinality). *On the MMANC with rate-R_h decoder assistance,*

$$C_{0,\mathrm{dec}}(R_h) \leq \log|\mathcal{A}| - \log|\mathcal{S}| + R_h \tag{5.21}$$

and

$$C_{0,\mathrm{dec}}(R_\mathrm{h}) \geq \frac{\log|\mathcal{A}|}{\log|\mathcal{S}|} \cdot R_\mathrm{h}$$

$$+ \left(1 - \frac{R_\mathrm{h}}{\log|\mathcal{S}|}\right) \cdot \max\left\{C_0, \frac{1}{2}\log\frac{|\mathcal{A}|}{|\mathcal{S}|}\right\}. \qquad (5.22)$$

These bounds also hold for $C_{0,\mathrm{enc}}(R_\mathrm{h})$ *and* $C_{0,\mathrm{both}}(R_\mathrm{h})$.

Theorem 5.8 has two corollaries. The first characterizes the zero-error helper capacity when the noise support "tessellates" the alphabet $\mathcal{A}$, and thus strengthens Theorem 5.5.

Corollary 5.9 (Special MMANCs). *On the MMANCs with* $C_0 = \log|\mathcal{A}| - \log|\mathcal{S}|$,

$$C_{0,\mathrm{both}}(R_\mathrm{h}) = C_{0,\mathrm{enc}}(R_\mathrm{h}) = C_{0,\mathrm{dec}}(R_\mathrm{h}) \qquad (5.23)$$

$$= \log|\mathcal{A}| - \log|\mathcal{S}| + R_\mathrm{h}. \qquad (5.24)$$

Proof of Corollary 5.9. Follows from Theorem 5.8 by substituting $\log|\mathcal{A}| - \log|\mathcal{S}|$ for C_0 on the RHS of (5.22) and noting that the result matches the RHS of (5.21). $\qquad\square$

When $\mathcal{S} = \mathcal{A}$, which implies that $C_0 = 0$ and hence that $C_0 = \log|\mathcal{A}| - \log|\mathcal{S}|$, the corollary recovers Theorem 5.5. But see Corollary 5.10 ahead for a stronger statement.

For another application of this corollary, consider the MMANCs with $|\mathcal{A}| = 4$ and with $\mathcal{S} = \{0,1\}$ or $\mathcal{S} = \{0,2\}$ (so $|\mathcal{S}| = 2$). In both cases $C_0 = 1$ $(= \log|\mathcal{A}| - \log|\mathcal{S}|)$, so the corollary yields that $C_{0,\mathrm{dec}}(R_\mathrm{h}) = C_{0,\mathrm{enc}}(R_\mathrm{h}) = 1 + R_\mathrm{h}$.

The second corollary to Theorem 5.8 provides a necessary and sufficient condition for the positivity of the zero-error helper capacity.

Corollary 5.10 (Positivity). *The following statements are equivalent:*

i) $C_{0,\mathrm{dec}}(R_\mathrm{h}) = 0$;

ii) $C_{0,\mathrm{enc}}(R_\mathrm{h}) = 0$;

iii) $C_{0,\mathrm{both}}(R_\mathrm{h}) = 0$;

iv) $C_{\mathrm{0F,dec}}(R_\mathrm{h}) = 0$;

v) $C_{0F,\text{enc}}(R_h) = 0$;

vi) $C_{0F,\text{both}}(R_h) = 0$;

vii) $R_h = 0$ *and* $\mathcal{S} = \mathcal{A}$.

Proof of Corollary 5.10. The equivalence of iv), v), vi), and vii) follows from Theorem 5.6 on feedback. The implications vii) $\implies$ i), ii), and iii) follow from (5.21) and the analogous result for $C_{0,\text{enc}}(R_h)$ and $C_{0,\text{both}}(R_h)$; the implication i) $\implies$ vii) follows from (5.22) as follows:

$$C_{0,\text{dec}}(R_h) \geq \frac{\log |\mathcal{A}|}{\log |\mathcal{S}|} \cdot R_h + \left(1 - \frac{R_h}{\log |\mathcal{S}|}\right) \cdot \frac{1}{2} \log \frac{|\mathcal{A}|}{|\mathcal{S}|} \tag{5.25}$$

$$= \frac{1}{2} \log \frac{|\mathcal{A}|}{|\mathcal{S}|} + \left(\frac{\log |\mathcal{A}|}{2 \log |\mathcal{S}|} + \frac{1}{2}\right) \cdot R_h \tag{5.26}$$

$$\geq \frac{1}{2} \log \frac{|\mathcal{A}|}{|\mathcal{S}|} + R_h; \tag{5.27}$$

the proof that ii) $\implies$ vii) and iii) $\implies$ vii) is similar. $\qquad\square$

The above theorems and corollaries have some noteworthy implications:

Remark 5.11 (Benefit of assistance). Assistance can increase the zero-error capacity by more than its rate. Even zero-rate assistance can increase the zero-error capacity: on the Pentagon channel, it raises the zero-error capacity from Lovász's $\frac{1}{2} \log 5$ to $\log \frac{5}{2}$, i.e., to C_{0F} (Corollary 5.3); on the Triangle channel, it raises the zero-error capacity from zero to $\log \frac{3}{2}$, which even exceeds C_{0F} (the latter being zero for this channel). $\hfill\triangle$

Remark 5.12. Thanks to Theorem 5.7, the zero-error capacity with a helper can sometimes be determined even if it is unknown in the absence of help, e.g., for the Heptagon channel. $\hfill\triangle$

Remark 5.13 (Less than one bit). As on the Gel'fand–Pinsker channel with feedback [50], in all scenarios, the zero-error capacity can be positive yet smaller than 1 bit. This is not the case in the absence of assistance. $\hfill\triangle$

In the following subsections, we present the proofs for the above results.

5.4 Feedback Link Present

In this section, we study the zero-error feedback capacity with a helper and prove Theorem 5.6. To this end, we need the following lemma, stating that feedback does not increase the Shannon capacity on the MMANC with a helper.

Lemma 5.14. *On the MMANC with feedback and rate-$\mathsf{R_h}$ assistance, the Shannon capacity is given by*

$$\mathsf{C}_{\mathrm{F,both}}(\mathsf{R_h}) = \mathsf{C}_{\mathrm{F,enc}}(\mathsf{R_h}) = \mathsf{C}_{\mathrm{F,dec}}(\mathsf{R_h}) \tag{5.28}$$

$$= \log|\mathcal{A}| - \left\{ H(Q_Z) - \mathsf{R_h} \right\}^{+}. \tag{5.29}$$

Proof. In light of Theorem 2.1, which establishes that the RHS of (5.29) can be achieved without feedback, we only need to prove a converse for $\mathsf{C}_{\mathrm{F,both}}(\mathsf{R_h})$. We assume $\mathsf{R_h} \le H(Q_Z)$, because otherwise the result is obvious.

Consider a message M that is drawn equiprobably from the message set $\mathcal{M}$. For any sequence of coding schemes of rate R with rate-$\mathsf{R_h}$ assistance and vanishing probabilities of error,

$$\begin{aligned}
\log|\mathcal{M}| &= H(M) & (5.30)\\
&= I(M; \mathbf{Y}, T) + H(M|\mathbf{Y}, T) & (5.31)\\
&\le I(M; \mathbf{Y}, T) + n\delta_n & (5.32)\\
&= I(M; \mathbf{Y}|T) + n\delta_n & (5.33)\\
&= H(\mathbf{Y}|T) - H(\mathbf{Y}|M, T) + n\delta_n & (5.34)\\
&\le H(\mathbf{Y}) - H(\mathbf{Y}|M, T) + n\delta_n & (5.35)\\
&\le H(\mathbf{Y}) - H(\mathbf{Z}|M, T) + n\delta_n & (5.36)\\
&= H(\mathbf{Y}) - H(\mathbf{Z}|T) + n\delta_n & (5.37)\\
&= H(\mathbf{Y}) - H(\mathbf{Z}) + I(\mathbf{Z}; T) + n\delta_n & (5.38)\\
&\le H(\mathbf{Y}) - H(\mathbf{Z}) + \log|\mathcal{T}| + n\delta_n & (5.39)\\
&\le n\log|\mathcal{A}| - nH(Q_Z) + \log|\mathcal{T}| + n\delta_n, & (5.40)
\end{aligned}$$

where (5.32) holds for some $\{\delta_n\}$ tending to zero by Fano's inequality; (5.33) and (5.37) hold because T is a function of $\mathbf{Z}$, so $(\mathbf{Z}, T)$ is independent of M; and (5.36) holds because, in the presence of feedback and help, $\mathbf{Z}$ is a function of $(\mathbf{Y}, M, T)$ namely $Z_i = Y_i \ominus f_i(M, T, Y^{i-1})$ for $i \in [n]$. Dividing the inequalities by n and letting n tend to infinity establishes the converse. $\qquad\square$

Now we move on to prove Theorem 5.6.

Converse Part. Consider the bi-terminal assistance.

If $\tilde{Q}_{Y|X}$ is any auxiliary MMANC over $\mathcal{A}$ of noise PMF $\tilde{Q}_Z \in \mathcal{P}(\mathcal{A})$ that is absolutely continuous with respect to Q_Z (i.e., whose support is contained in $\mathcal{S}$, denoted $\tilde{Q}_Z \ll Q_Z$), then its Shannon capacity with feedback and bi-terminal assistance $\tilde{C}_{F,both}(R_h)$ forms an upper bound on $C_{0F,both}(R_h)$, because any error-free coding scheme for the original channel is also error-free on the auxiliary channel. Indeed, for any $\mathbf{y} \in \mathcal{A}^n$ and $t \in \mathcal{T}$, the absolute continuity hypothesis implies that

$$\left(\tilde{Q}^n_{Y|X}(\mathbf{y}|\mathbf{x}(m,\mathbf{y})) > 0 \right) \implies \left(Q^n_{Y|X}(\mathbf{y}|\mathbf{x}(m,\mathbf{y})) > 0 \right) \tag{5.41}$$

so if a message m is compatible with $(\mathbf{y}, t)$ on the auxiliary channel (in the sense that $\tilde{Q}^n_{Y|X}(\mathbf{y}|\mathbf{x}(m,\mathbf{y})) > 0$ and $h(\mathbf{y} \ominus \mathbf{x}(m,\mathbf{y})) = t$), then it is also compatible with $(\mathbf{y}, t)$ on the original channel.

Therefore, upon minimizing over the choice of $\tilde{Q}_{Y|X}$ to get the tightest bound,

$$C_{0F,both}(R_h) \leq \min_{\tilde{Q}_Z \,:\, \tilde{Q}_Z \ll Q_Z} \tilde{C}_{F,both}(R_h) \tag{5.42}$$

$$= \min_{\tilde{Q}_Z \,:\, \tilde{Q}_Z \ll Q_Z} \left\{ \log|\mathcal{A}| - \left\{ H(\tilde{Q}_Z) - R_h \right\}^+ \right\} \tag{5.43}$$

$$= \log|\mathcal{A}| - \left\{ \log|\mathcal{S}| - R_h \right\}^+, \tag{5.44}$$

where (5.43) follows from Lemma 5.14, and (5.44) holds because, subject to a support constraint, the uniform PMF maximizes the entropy.

Direct Part. We distinguish three different cases.

- Case 1: $R_h = \log|\mathcal{S}|$.

In this case, feedback is unnecessary. The codebook comprises all the distinct sequences in $\mathcal{A}^n$. Using $\lceil n \log|\mathcal{S}| \rceil$ bits, the helper describes the noise sequence $\mathbf{Z}$ precisely. The decoder (resp. encoder) subtracts the noise from the received sequence (resp. from the codeword to be transmitted), so the codeword and the message can be received error-free. This establishes the achievability of $\log|\mathcal{A}|$ bits per channel use.

- Case 2: $R_h = 0$.

A two-phase coding scheme is proposed.

In Phase 1, we follow the construction (for a uniform input distribution) in Shannon's proof of Theorem 5.2 in [47], where the encoder sequentially reduces the decoder's ambiguity. In the i-th channel use, thanks to the feedback, the encoder reconstructs the list of messages compatible with Y^{i-1} and evenly assigns them to different input symbols (in a manner agreed upon with the decoder prior to transmission). Only $|\mathcal{S}|$ of the $|\mathcal{A}|$ input symbols are compatible with Y_i, and the number of compatible messages is reduced by a factor of roughly $\frac{|\mathcal{S}|}{|\mathcal{A}|}$. More precisely, Shannon showed that if $|\mathcal{M}| = \lfloor (\frac{|\mathcal{S}|}{|\mathcal{A}|})^{-n} \rfloor$, then after n channel uses, the number of compatible messages is at most $|\mathcal{A}|^2$.

The final ambiguity is removed in Phase 2, where the helper comes into play. Since the messages that are compatible with the outputs from Phase 1 are known to the encoder, and since their number does not exceed $|\mathcal{A}|^2$, the encoder can inform the decoder which compatible message was sent in two additional clean channel uses. To clean these two channel uses, the helper informs the decoder (resp. encoder) of the exact value of $Z_{n+1}^{n+2} \in \mathcal{S}^2$ and the decoder (resp. encoder) subtracts the noise after (resp. before) the transmission.

The rate of help is therefore

$$\lim_{n \to \infty} \frac{1}{n+2} \log |\mathcal{S}|^2 = 0 \tag{5.45}$$

and the transmission rate

$$\lim_{n \to \infty} \frac{1}{n+2} \log |\mathcal{M}| = \lim_{n \to \infty} \frac{\log \lfloor (\frac{|\mathcal{S}|}{|\mathcal{A}|})^{-n} \rfloor}{n+2} = \log \frac{|\mathcal{A}|}{|\mathcal{S}|}. \tag{5.46}$$

- Case 3: $0 < \mathsf{R}_\mathsf{h} < \log |\mathcal{S}|$.

We divide the transmission block into two parts of relative length $\frac{\mathsf{R}_\mathsf{h}}{\log |\mathcal{S}|}$ and $1 - \frac{\mathsf{R}_\mathsf{h}}{\log |\mathcal{S}|}$. We then apply the aforementioned coding schemes for helper rates of $\log |\mathcal{S}|$ and zero, respectively. The transmission is error-free because the probabilities of error in both blocks are exactly zero. The total rate achieved by this time-sharing scheme is

$$\frac{\mathsf{R}_\mathsf{h} \log |\mathcal{A}|}{\log |\mathcal{S}|} + \left(1 - \frac{\mathsf{R}_\mathsf{h}}{\log |\mathcal{S}|}\right) \left(\log |\mathcal{A}| - \log |\mathcal{S}|\right)$$
$$= \log |\mathcal{A}| - \log |\mathcal{S}| + \mathsf{R}_\mathsf{h}. \tag{5.47}$$

$\square$

5.5　Feedback Link Absent

In this section, we provide proofs pertaining to the zero-error helper capacity in the absence of feedback.

5.5.1　Prime Cardinality

We begin with the case where $|\mathcal{A}|$ is a prime and prove Theorem 5.7. Denote $|\mathcal{A}|$ by p, and denote the cardinality-p finite field $\mathbb{F}_p$ and identify it with the set $\mathbb{Z}_p = \{0, \ldots, p-1\}$ with mod-p arithmetic.

Since feedback cannot hurt, it follows from Theorem 5.6 that we only need to prove the direct part. This is trivial unless $|\mathcal{S}| < |\mathcal{A}|$, which we proceed to assume. We first focus on decoder assistance.

- Case 1: $R_h = \log|\mathcal{S}|$.

  The achievability in this case is as in the proof of Theorem 5.6, where the feedback link is ignored.

- Case 2: $R_h = 0$.

  We will construct a sequence of blocklength-n codebooks of rate $\left(\log \frac{|\mathcal{A}|}{|\mathcal{S}|} - \epsilon_n\right)$ that can be decoded error-free utilizing rate-ϵ'_n decoder assistance, for some $\{\epsilon_n\}$ and $\{\epsilon'_n\}$ tending to zero.

  The codes we construct have two key properties. The first is that they are L-*list-decodable* [51] where $\mathsf{L} \in \mathbb{Z}^+$ grows subexponentially with n. That is, every $\mathbf{y} \in \mathcal{A}^n$ is compatible with at most L messages. Only in such case, the decoder's ambiguity could be eliminated with a sublinear number of bits. Elias [51] established the existence of such codebooks of rate $\log \frac{|\mathcal{A}|}{|\mathcal{S}|} - \Theta(\mathsf{L}^{-1})$. But this is not enough, because, in the absence of feedback, neither the transmitter nor the helper can determine the list facing the decoder. This is where the second property comes in: To overcome this issue and enable the helper to remove the ambiguity, we shall impose a linear structure on the code, and this is where the assumption that $|\mathcal{A}|$ is a prime will be essential: it will allow us to view $\mathcal{A}$ as a field.

  The existence of *linear* L-list-decodable codes can be established using a variation on a theme by Elias [51] using tools that were used successfully in the analysis of random linear codes (e.g. [52]–[54]).

Lemma 5.15. *Consider a MMANC with* $|\mathcal{A}| = p$, *where p is prime.*

Given $\mathsf{L} \in \mathbb{Z}^+$, *define*

$$\mathsf{R}_\mathsf{L} = \max\left\{0,\ \log\frac{|\mathcal{A}|}{|\mathcal{S}|} - \frac{\log^2 |\mathcal{A}|}{\log(\mathsf{L}+1)}\right\}. \tag{5.48}$$

Then, for any $n \in \mathbb{Z}^+$, there exists a blocklength-n linear code over the field $\mathbb{F}_p$ of rate $\frac{\log|\mathcal{A}|}{n}\left\lfloor\frac{n\mathsf{R}_\mathsf{L}}{\log|\mathcal{A}|}\right\rfloor$ that is L-list-decodable.

Proof. Assume $\mathsf{R}_\mathsf{L} > 0$ (because otherwise there is nothing to prove). Given some blocklength n, let the message set be $\mathcal{M} = \mathbb{F}_p^k$, with $k \in \mathbb{Z}^+$ to be specified later. A generic message $\mathbf{m} \in \mathcal{M}$ is thus represented by a k-vector, and the transmission rate is k/n in base-p logarithm, or $(k/n)\log_2 p$ bits per channel use.

Pick a random $(n \times k)$-matrix $\mathbb{A}$ whose entries are drawn IID equiprobably from $\mathbb{F}_p$, and consider the encoding rule $\mathbf{m} \mapsto \mathbf{X}(\mathbf{m}) = \mathbb{A}\mathbf{m}$. Let $\mathcal{C}$ be the random linear code (multiset) it induces. This encoding rule maps any ℓ linearly independent messages to independent codewords, each having IID equiprobable random components.

Among any $(\mathsf{L}+1)$ messages, at least $\ell \triangleq \lceil\log_p(\mathsf{L}+1)\rceil$ are linearly independent, so the probability that there exists some $\mathbf{y} \in \mathcal{A}^n$ compatible with $(\mathsf{L}+1)$ messages is upper bounded by the probability that there exists some $\mathbf{y} \in \mathcal{A}^n$ that is compatible with ℓ linearly independent messages. The latter, by the Union Bound, is strictly smaller than

$$\sum_{\substack{\mathbf{y}\in\mathcal{A}^n}} \sum_{\substack{\mathbf{m}_1,\ldots,\mathbf{m}_\ell\in\mathcal{M}\\ \text{linearly independent}}} \mathbb{P}\bigcap_{i=1}^{\ell}\bigcap_{j=1}^{n}\left[X_j(\mathbf{m}_i)\in\mathcal{X}_{y_j}\right]$$

$$= \sum_{\substack{\mathbf{y}\in\mathcal{A}^n}} \sum_{\substack{\mathbf{m}_1,\ldots,\mathbf{m}_\ell\in\mathcal{M}\\ \text{linearly independent}}} \left(\frac{|\mathcal{S}|^n}{|\mathcal{A}|^n}\right)^\ell \tag{5.49}$$

$$\leq |\mathcal{A}|^n \binom{|\mathcal{M}|}{\ell}\left(\frac{|\mathcal{S}|}{|\mathcal{A}|}\right)^{n\ell} \tag{5.50}$$

$$\leq |\mathcal{A}|^n p^{k\ell}\left(\frac{|\mathcal{S}|}{|\mathcal{A}|}\right)^{n\ell} \tag{5.51}$$

$$\leq \left(|\mathcal{A}|\, p^{\left(\ell\log_p\frac{|\mathcal{A}|}{|\mathcal{S}|}-1\right)}\left(\frac{|\mathcal{S}|}{|\mathcal{A}|}\right)^\ell\right)^n \tag{5.52}$$

$$= 1, \tag{5.53}$$

where (5.49) holds because—when $\mathbf{m}_1,\ldots,\mathbf{m}_\ell$ are linearly independent—the codewords $\mathbf{X}(\mathbf{m}_1),\ldots,\mathbf{X}(\mathbf{m}_\ell)$ are independent, each having IID equiprobably distributed components, and because $|\mathcal{X}_y| = |\mathcal{S}|$ for every $y \in \mathcal{A}$; and in (5.52) we choose $k = \lfloor n\big(\log_p \frac{|\mathcal{A}|}{|\mathcal{S}|} - \frac{1}{\log_p(L+1)}\big)\rfloor$, so

$$k\ell \leq n\left(\log_p \frac{|\mathcal{A}|}{|\mathcal{S}|} - \frac{1}{\log_p(L+1)}\right)\ell \qquad (5.54)$$

$$\leq n\left(\ell\log_p \frac{|\mathcal{A}|}{|\mathcal{S}|} - 1\right). \qquad (5.55)$$

Hence, with positive probability, the random linear code is L-list-decodable. The lemma then follows by noting that, in bits, the rate is $\frac{k}{n}\cdot\log|\mathcal{A}|$. $\square$

We now use Lemma 5.15 to complete the proof of Theorem 5.7 for the case of $\mathsf{R}_\mathrm{h} = 0$. Let $\{\mathsf{L}_n\}$ be a sequence of positive integers tending to infinity subexponentially, e.g., $\mathsf{L}_n = \Theta(n)$. The lemma implies that, for every blocklength n, there exists a linear code $\mathcal{C}_n$ of rate $\frac{\log|\mathcal{A}|}{n}\lfloor\frac{n\mathsf{R}_{\mathsf{L}_n}}{\log|\mathcal{A}|}\rfloor$ that is L_n-list-decodable. A minor annoyance is that the codewords in $\mathcal{C}_n$ need not be distinct. To overcome this, we consider the code $\mathcal{C}'_n \subseteq \mathbb{F}_p^n$ comprising all the distinct elements in $\mathcal{C}_n$. Note that (i) $\mathcal{C}'_n$ is a subgroup of $\mathbb{F}_p^n$; (ii) $\mathcal{C}'_n$ is L_n-list-decodable; and (iii) its cardinality satisfies

$$|\mathcal{C}'_n| \geq \frac{|\mathcal{C}_n|}{\mathsf{L}_n} \qquad (5.56)$$

(because $\mathcal{C}_n$, being L_n-list-decodable, contains no codeword more than L_n times). This latter property and the fact that $\{\mathsf{L}_n\}$ is subexponential imply that $\{\mathcal{C}'_n\}$ has the desired rate:

$$\lim_{n\to\infty}\frac{1}{n}\log|\mathcal{C}'_n| = \lim_{n\to\infty}\frac{1}{n}\log|\mathcal{C}_n| \qquad (5.57)$$

$$= \lim_{n\to\infty}\mathsf{R}_{\mathsf{L}_n} \qquad (5.58)$$

$$= \lim_{n\to\infty}\log\frac{|\mathcal{A}|}{|\mathcal{S}|} - \frac{\log^2|\mathcal{A}|}{\log(\mathsf{L}_n+1)} \qquad (5.59)$$

$$= \log\frac{|\mathcal{A}|}{|\mathcal{S}|}, \qquad (5.60)$$

where (5.57) follows from (5.56) and the fact that $\{\mathsf{L}_n\}$ is subexponential; and (5.60) holds because $\{\mathsf{L}_n\}$ tends to infinity.

We next show that—although the helper is incognizant of the list of messages that are compatible with the received sequence—a $\lceil \log \mathsf{L}_n \rceil$-bit description of the noise sequence (which is of zero rate as L_n is subexponential in the blocklength n) suffices to guarantee zero-error transmission of the codebook $\mathcal{C}'_n$. To this end, we propose the following helper. To simplify its description, we drop the subscript n.

For $\mathbf{z}, \mathbf{z}' \in \mathcal{S}^n$, let us write $\mathbf{z} \sim \mathbf{z}'$ if their componentwise difference is in $\mathcal{C}'$, i.e.,

$$\left(\mathbf{z} \sim \mathbf{z}'\right) \iff \left(\mathbf{z} \ominus \mathbf{z}' \in \mathcal{C}'\right), \quad \mathbf{z}, \mathbf{z}' \in \mathcal{S}^n. \tag{5.61}$$

Since $\mathcal{C}'$ is a subgroup of $\mathbb{F}_p^n$, this relation is an equivalence relation, and $\mathbf{z} \sim \mathbf{z}'$, i.e., $\mathbf{z}$ and $\mathbf{z}'$ are equivalent, if, and only if, $\mathbf{z}$ and $\mathbf{z}'$ belong to the same coset of $\mathcal{C}'$. We shall use $[\mathbf{z}] \subseteq \mathcal{S}^n$ to denote the equivalence class containing $\mathbf{z}$.

Our proposed helper assigns labels (descriptions) only to noise sequences in $\mathcal{S}^n$, and it does so in such a way that, unless identical, *equivalent noise sequences are assigned differing labels.* Such a helper leads to zero errors, because if $\mathbf{x} \in \mathcal{C}'$ is transmitted and $\mathbf{x} \oplus \mathbf{z}$ is received (where $\mathbf{z} \in \mathcal{S}^n$), then the decoder can confuse $\mathbf{x}$ with some $\mathbf{x}'$ only if: (i) $\mathbf{x}'$ is also a codeword; (ii) $\mathbf{x} \oplus \mathbf{z} = \mathbf{x}' \oplus \mathbf{z}'$ for some $\mathbf{z}' \in \mathcal{S}^n$; and (iii) $\mathbf{z}$ and $\mathbf{z}'$ have the same label. The former two conditions imply that $\mathbf{z} \sim \mathbf{z}'$, and hence that $\mathbf{z}$ and $\mathbf{z}'$ are identical or else of differing labels. The third condition then implies that they are, in fact, identical, so $\mathbf{x}'$ equals $\mathbf{x}$.

It remains to verify that we can find a labeling rule as above with at most L different labels. This will follow once we show that, for every $\mathbf{z} \in \mathcal{S}^n$,

$$\left| [\mathbf{z}] \right| \leq \mathsf{L}. \tag{5.62}$$

To establish (5.62), we note that the L-list-decodability property of $\mathcal{C}'$, namely

$$\mathsf{L} \geq \left| \left(\mathbf{y} \ominus \mathcal{C}'\right) \cap \mathcal{S}^n \right|, \quad \mathbf{y} \in \mathcal{A}^n \tag{5.63}$$

is equivalent (because $\mathcal{C}'$ is a subgroup of $\mathbb{Z}_p^n$) to

$$\mathsf{L} \geq \left| \left(\mathbf{y} \oplus \mathcal{C}'\right) \cap \mathcal{S}^n \right|, \quad \mathbf{y} \in \mathcal{A}^n, \tag{5.64}$$

i.e., to every coset of $\mathcal{C}'$ intersecting $\mathcal{S}^n$ in at most L points. This establishes (5.62) and concludes the achievability proof.

- Case 3: $0 < \mathsf{R_h} < \log |\mathcal{S}|$.

  Follows from time sharing.

The case with encoder assistance is essentially identical. If $\mathsf{R_h} = \log|\mathcal{S}|$, the rate $\log|\mathcal{A}|$ is achievable as in the proof of Theorem 5.6. If $\mathsf{R_h} = 0$, the relation

$$\mathsf{C}_{0,\mathrm{enc}}(0) \geq \mathsf{C}_{0,\mathrm{dec}}(0) \tag{5.65}$$

holds because, in the presence of encoder assistance, any zero-rate help to the encoder can be conveyed to the decoder with negligible extra help and negligible loss in rate: the encoder simply appends a frame to convey the help, with the frame being of sublinear length (because the help to be conveyed is of zero rate); it requests that the helper provide it with a precise description of the noise affecting the frame (with the extra help being negligible because the frame is short); and it subtracts that noise from the transmission in that frame so as to render it noise free. For intermediate values of $\mathsf{R_h}$, the achievability follows by time sharing. $\qquad\square$

Remark 5.16 (Gap to capacity vs L). By Lemma 5.15, as the number of labels L tends to infinity, it is possible to communicate error-free at transmission rates that converge to the zero-error helper capacity, with the gap to capacity decaying in L like $\mathcal{O}(1/\log \mathsf{L})$. Although irrelevant to the computation of the capacity, it might be interesting to investigate whether the gap to capacity can decay faster in L. $\qquad\triangle$

Remark 5.17 (L-list-decodability and (ℓ, L) recoverability). On the MMANC, L-list-decodability is related to (zero-error) list-recoverability [54]: Given $\ell, \mathsf{L} \in \mathbb{Z}^+$ and a finite set $\mathcal{X}$, a codebook $\mathcal{C} \subseteq \mathcal{X}^n$ is (ℓ, L)-*list-recoverable* if for any collection of n subsets $\mathcal{S}_1, \mathcal{S}_2, \ldots, \mathcal{S}_n$ of $\mathcal{X}$, each of which has no more than ℓ elements,

$$\left| \mathcal{C} \cap \left(\mathcal{S}_1 \times \cdots \times \mathcal{S}_n \right) \right| \leq \mathsf{L}. \tag{5.66}$$

On the MMANC, $(|\mathcal{S}|, \mathsf{L})$-list-recoverability implies L-list-decodability, because, given any output sequence $\mathbf{y} \in \mathcal{A}^n$, we can substitute $y_i \ominus \mathcal{S}$ for each $\mathcal{S}_i$ in (5.66) to recover L-list-decodability. $\qquad\triangle$

Remark 5.18. If instead of defining $\mathsf{R_L}$ as in (5.48), we defined it as

$$\log \frac{|\mathcal{A}|}{|\mathcal{S}|} - \frac{|\mathcal{S}| \cdot \log^2 |\mathcal{A}|}{\log(\mathsf{L}+1)}, \tag{5.67}$$

then the resulting weaker version of Lemma 5.15, while still sufficient for our purposes, could have been recovered from the literature on (ℓ, L)-list-recoverability, specifically from the result that a random linear code of such rate is $(|\mathcal{S}|,\mathsf{L})$-list-recoverable with high probability [52, 55] and, a fortiori, with positive probability. $\triangle$

Remark 5.19. The factor of $|\mathcal{S}|$ in the numerator of the second term on the RHS of (5.67), which is absent from (5.48), can be improved for large $|\mathcal{A}|$ using [56, Theorem 5.1] and [55]. $\triangle$

5.5.2 General Case

We now turn to the general case where $|\mathcal{A}|$ need not be prime and establish Theorem 5.8.

As in the proof of Theorem 5.7, we only need to prove the direct part, and we focus on decoder assistance, so our goal is to establish that

$$
\mathsf{C}_{0,\mathrm{dec}}(\mathsf{R}_\mathrm{h}) \geq \frac{\log |\mathcal{A}|}{\log |\mathcal{S}|} \cdot \mathsf{R}_\mathrm{h}
$$

$$
+ \left(1 - \frac{\mathsf{R}_\mathrm{h}}{\log |\mathcal{S}|}\right) \cdot \max\left\{\mathsf{C}_0, \frac{1}{2}\log \frac{|\mathcal{A}|}{|\mathcal{S}|}\right\}. \tag{5.68}
$$

The achievability for encoder assistance will then follow as in the proof of Theorem 5.7. To establish (5.68), we propose the following coding scheme based on time sharing.

- Case 1: $\mathsf{R}_\mathrm{h} = \log |\mathcal{S}|$.

  That $\mathsf{C}_{0,\mathrm{dec}}(\mathsf{R}_\mathrm{h}) \geq \log |\mathcal{A}|$ follows from the proof for Theorem 5.6, where the feedback link is not utilized.

- Case 2: $\mathsf{R}_\mathrm{h} = 0$.

  That $\mathsf{C}_{0,\mathrm{dec}}(0) \geq \mathsf{C}_0$ is obvious because help cannot hurt.

  To show that

$$
\mathsf{C}_{0,\mathrm{dec}}(0) \geq \frac{1}{2}\log \frac{|\mathcal{A}|}{|\mathcal{S}|}, \tag{5.69}
$$

we first introduce some notation. Given a codebook $\mathcal{C} \subseteq \mathcal{A}^n$ and a noise sequence $\mathbf{z} \in \mathcal{S}^n$, let $\mathcal{F}_\mathbf{z}(\mathcal{C})$—or $\mathcal{F}_\mathbf{z}$ for short—be the confusion set of $\mathbf{z}$

comprising the noise sequences confusable with $\mathbf{z}$:

$$\mathcal{F}_{\mathbf{z}}(\mathcal{C}) = \left\{ \mathbf{z}' \in \mathcal{S}^n : \exists\, \mathbf{x}, \mathbf{x}' \in \mathcal{C} \text{ s.t. } \mathbf{x} \neq \mathbf{x}' \right.$$

$$\left. \text{and } \mathbf{z} \oplus \mathbf{x} = \mathbf{z}' \oplus \mathbf{x}' \right\} \tag{5.70}$$

$$= \left\{ \mathbf{z}' \in \mathcal{S}^n : \mathbf{z}' \ominus \mathbf{z} \in (\mathcal{C} \ominus \mathcal{C})^{\star} \right\}. \tag{5.71}$$

The proposed helper assigns confusable noise sequences different labels: if $\mathbf{z}' \in \mathcal{F}_{\mathbf{z}}$, then the labels assigned to $\mathbf{z}$ and $\mathbf{z}'$ are different. This guarantees error-free recovery of the noise sequence and hence, if the code has no repeating codewords, also of the transmitted message.

As we next argue, the number of different labels required is at most

$$\max_{\mathbf{z} \in \mathcal{S}^n} \left| \mathcal{F}_{\mathbf{z}} \right| + 1. \tag{5.72}$$

Indeed, the number of required labels is the chromatic number of the confusion graph of the noise sequences, i.e., the undirected graph with vertices $\mathcal{S}^n$ and with $\mathbf{z}$ connected to $\mathbf{z}'$ if $\mathbf{z}' \in \mathcal{F}_{\mathbf{z}}$. This graph is well-defined because

$$\left(\mathbf{z} \ominus \mathbf{z}' \in (\mathcal{C} \ominus \mathcal{C})^{\star} \right) \iff \left(\mathbf{z}' \ominus \mathbf{z} \in (\mathcal{C} \ominus \mathcal{C})^{\star} \right). \tag{5.73}$$

The degree of a vertex $\mathbf{z}$ in this graph is $\left| \mathcal{F}_{\mathbf{z}} \right|$, and our claimed upper bound on the number of required labels follows from the fact that the chromatic number of any graph is upper bounded by its maximum degree plus 1.

It remains to establish the existence of a code with no repeating codewords that induce small confusion sets. This is established by the following lemma, whose proof is postponed to Appendix C.2.

Lemma 5.20. *On the MMANC, let* $\mathsf{L} > 3$ *be a positive integer, and define*

$$\mathsf{R}_{\mathsf{L}} = \max\left\{ 0, \frac{1}{2} \log \frac{|\mathcal{A}|}{|\mathcal{S}|} - \frac{\log |\mathcal{S}|}{2 \log_3 \mathsf{L}} \right\}. \tag{5.74}$$

Then for any $n \in \mathbb{Z}^+$, *there exists a codebook of cardinality* $\lfloor 2^{n\mathsf{R}_{\mathsf{L}}} \rfloor$, *of differing codewords, and for which* $\max_{\mathbf{z} \in \mathcal{S}^n} \left| \mathcal{F}_{\mathbf{z}} \right| \leq \mathsf{L} - 1$.

With the aid of the lemma, we can conclude the achievability for zero-rate help using arguments similar to those we used in the proof of Theorem 5.7: Consider a sequence of blocklength-n codebooks whose existence is guaranteed by Lemma 5.20 when we substitute L_n for L, where $\{L_n\}$ tends to infinity subexponentially in n. With these codebooks and the proposed helper, transmission is error-free, the helping rate is zero, and the transmission rate approaches $\frac{1}{2}\log\frac{|\mathcal{A}|}{|\mathcal{S}|}$, thus proving (5.69).

- Case 3: $0 < R_\mathrm{h} < \log|\mathcal{S}|$.

Follows from time sharing, by dividing the transmission block into two parts of relative length $\frac{R_\mathrm{h}}{\log|\mathcal{S}|}$ and $1 - \frac{R_\mathrm{h}}{\log|\mathcal{S}|}$ and applying the aforementioned schemes. $\qquad\square$

Part II

Message Cognizance and Feedback

Chapter 6

Effects of Message-Cognizance and Feedback

6.1   Introduction

The state-dependent discrete memoryless channel (SD-DMC) is studied here in the presence of a benevolent, message-cognizant, rate-limited helper that observes the state sequence noncausally, produces a message-dependent rate-limited description of it, and provides said description to both encoder and decoder; see Fig 1.5.

Key is that the description may depend on the message that the encoder wishes to send, which affords the helper flexibility that may increase capacity. Also noteworthy is that assistance is provided to both the encoder and the decoder. Since the helper is cognizant of the message, it can, for example, embed information bits in the description and in this way convey them to the decoder error-free. This introduces tension between using the assistance bits to describe the state and using them to send data bits directly.

Unlike [15], which mostly focuses on causal help, and [12], which focuses on symbol-by-symbol help—our focus is on noncausal help. The state sequence is thus observed by the helper noncausally, and its description is provided to the encoder (and to the decoder) before transmission begins.

A scenario that might lead to our setting is one where a user wishes to download a movie, which is stored as a file on two servers. One of the servers, perhaps the one with the cleaner downstream channel to the user, takes upon itself the job of transmitting the file. The other server, however, need not be idle: it can assume the role of a helper that is cognizant of the transmitted message (file).

In the Gaussian version of this problem, the state corresponds to additive Gaussian noise; the encoder is power limited; and the alphabets are infinite. Our approach to this setting is thus geometric and does not resort to the Method of Types. The results in this setting are thus also accessible to readers who skip the sections on the SD-DMC.

For this setting, our results complement the recent results on message-oblivious helpers in [2,3,5] and in Part I of the thesis. Also related to our work is [8], which focuses mainly (but not exclusively) on the transmission of a random parameter rather than a message, with assistance being provided to the encoder only.

An interesting role is played in the Gaussian setting by feedback from the receiver to the encoder: Even if ideal, it does not increase capacity, but its presence allows one to achieve capacity even when the helper is incognizant of the transmitted message.

The remainder of this chapter is organized as follows: Section 6.2 introduces the problem setup on the SD-DMC and presents the main results, along with a discussion of its implications on various examples. Those on the Gaussian channel are presented in Section 6.3. Proofs for these two cases are discussed in Section 6.4 and Section 6.5, respectively.

6.2   SD-DMC: Main Result and Discussion

6.2.1   Problem Formulation

We consider an SD-DMC with finite input, output, and state alphabets $\mathcal{X}$, $\mathcal{Y}$, and $\mathcal{S}$, respectively. The states $\{S_k\}$ are drawn IID according to some PMF Q_S independently of the message M to be transmitted. Given input $x \in \mathcal{X}$ and state $s \in \mathcal{S}$, the probability of the output being $y \in \mathcal{Y}$ is $W(y|x,s)$.

A blocklength-n rate-R coding scheme with rate-R_h message-cognizant assistance comprises a message set

$$\mathcal{M} = \left\{1, \ldots, 2^{n\mathsf{R}}\right\}; \tag{6.1}$$

a description set

$$\mathcal{T} = \left\{1, \ldots, 2^{n\mathsf{R_h}}\right\}; \tag{6.2}$$

a helper that observes the state sequence S^n noncausally as well as the message M and produces a description, represented by the assistance function

$$\mathsf{h}\colon \mathcal{S}^n \times \mathcal{M} \to \mathcal{T} \tag{6.3a}$$

$$(s^n, m) \mapsto t; \tag{6.3b}$$

an encoder represented by the encoding function

$$\mathsf{f}\colon \mathcal{M} \times \mathcal{T} \to \mathcal{X}^n \tag{6.4a}$$

$$(m, t) \mapsto x^n; \tag{6.4b}$$

and a decoder that produces the decoded message using the decoding function

$$\phi\colon \mathcal{Y}^n \times \mathcal{T} \to \mathcal{M} \tag{6.5a}$$

$$(y^n, t) \mapsto \hat{m}. \tag{6.5b}$$

A rate R is said to be achievable if there exists a sequence of coding schemes as above (indexed by the blocklength) for which

$$\lim_{n \to \infty} \mathbb{P}[\hat{M} \neq M] = 0 \tag{6.6}$$

whenever M is drawn equiprobably from $\mathcal{M}$. The supremum of achievable rates is denoted $\mathsf{C}(\mathsf{R_h})$ and is the capacity we seek.

The feedback capacity is defined in an analogous way with the transmitted sequence x^n now having the form

$$x^n(m, t, y^{n-1}) = \left(x_1(m, t), x_2(m, t, y_1), \ldots, x_n(m, t, y^{n-1})\right).$$

It captures a scenario where, thanks to the feedback link from the channel output to the encoder, the time-k transmitted symbol may depend not only on the message m and on the help t but also on the previously-received symbols y^{k-1}.

Remark 6.1. The definition of rates in this part is slightly different from that in Part I of the thesis, c.f., (2.20) and (2.21). In particular, that of the assistance rate implies that a zero-rate helper is equivalent to no help. We made this change because, as we shall see, the effect of assistance on the Shannon capacity is continuous in the helping rate, so we follow the definition (6.1) and (6.2) to simplify the notation and the discussion. $\triangle$

6.2.2 Main Result

Our main result on the SD-DMC is a single-letter expression that equals the capacity both in the absence of feedback and in its presence. Before formally stating the result, we introduce this expression and some of its properties.

Definition 6.2. Define

$$\mathsf{C}^{(\mathrm{I})}(\mathsf{R_h}) \triangleq \max_{\substack{Q_V,Q_{U|S,V},Q_{X|U,V}:\\ I(U;S|V)\leq \mathsf{R_h}}} I(U;Y|V) - I(U;S|V) + \mathsf{R_h} \qquad (6.7)$$

where U and V are auxiliary random variables taking values in finite sets $\mathcal{U}$ and $\mathcal{V}$, respectively, and where the mutual information is computed with respect to the joint distribution

$$Q_V(v)\,Q_S(s)\,Q_{U|S,V}(u|s,v)\,Q_{X|U,V}(x|u,v)\,W(y|x,s). \qquad (6.8)$$

Proposition 6.3. *Without reducing the maximum in* (6.7), $Q_{X|U,V}$ *can be chosen to be deterministic, and the cardinalities of* V *and* $\mathcal{U}$ *can be restricted to* $|\mathcal{V}| \leq 3$ *and* $|\mathcal{U}| \leq |\mathcal{X}| \cdot |\mathcal{S}| + 1$ *respectively.*

Proof. See Appendix D.1. $\square$

Proposition 6.4. *An equivalent expression for* $\mathsf{C}^{(\mathrm{I})}(\mathsf{R_h})$ *is*

$$\mathsf{C}^{(\mathrm{I})}(\mathsf{R_h}) = \max_{\mathsf{R_0}\geq 0} \max_{\substack{Q_V,Q_{U|S,V},Q_{X|U,V}:\\ I(U;S|V)\leq \mathsf{R_h}-\mathsf{R_0}}} I(U;Y|V) + \mathsf{R_0}. \qquad (6.9)$$

Proof. One can swap the order of the two maximizations on the RHS of (6.9) to rewrite it as

$$\max_{\substack{Q_V,Q_{U|S,V},Q_{X|U,V}:\\ I(U;S|V)\leq \mathsf{R_h}}} \max_{0\leq \mathsf{R_0}\leq \mathsf{R_h}-I(U;S|V)} I(U;Y|V) + \mathsf{R_0}. \qquad (6.10)$$

Clearly, the inner maximum is achieved by $R_0 = R_h - I(U; S|V)$, yielding the expression on the RHS of (6.7). $\qquad\square$

Theorem 6.5. *The capacity $C(R_h)$ of the SD-DMC with noncausal rate-R_h message-cognizant assistance that is provided to both encoder and decoder is*

$$C(R_h) = C^{(I)}(R_h). \tag{6.11}$$

This remains the capacity also when a feedback link from the receiver to the encoder is added.

Proof. The proof of the direct part, which does not utilize the feedback link, is provided in Section 6.4.1. The converse, which is valid also in the presence of feedback, is provided in Section 6.4.2. $\qquad\square$

Remark 6.6. We defined the capacity under the average-probability-of-error criterion (6.6), but it is unaltered if we adopt the maximal-probability-of-error criterion. The only extra step in the code construction is to discard half the messages, specifically those of the highest probability of error. $\qquad\triangle$

6.2.3   Discussion

To better understand Theorem 6.5, we discuss below some special cases.

Absence of help.   As a sanity check, consider $R_h = 0$, corresponding to the absence of help. We have

$$C(0) = \max_{Q_V, Q_{U,X|V}} I(U; Y|V) \tag{6.12}$$

$$= \max_{Q_{U,X}} I(U; Y) \tag{6.13}$$

$$= \max_{Q_X} I(X; Y), \tag{6.14}$$

where the last step follows by noting that because S is independent of (U, X), the Markov chain $U \, \multimap \, X \, \multimap \, Y$ holds, and the Data Processing Inequality can be applied.

Useless channel: $W(y|x,s)$ does not depend on x. For such a channel, it always holds that $I(U;Y|V) \leq I(U;S|V)$, so $\mathsf{C}^{(\mathrm{I})}(\mathsf{R_h}) = \mathsf{R_h}$ (achieved by choosing U to be independent of S). Operationally, this capacity can be achieved by having the helper forgo describing the state sequence and having it only send information bits to the receiver.

Large assistance rate: $\mathsf{R_h} \geq H(S)$. The condition $I(U;S|V) \leq \mathsf{R_h}$ is always satisfied, so we can write $\mathsf{C}^{(\mathrm{I})}(\mathsf{R_h})$ as

$$\mathsf{C}^{(\mathrm{I})}(\mathsf{R_h}) = \max_{Q_V,Q_{U|S,V},Q_{X|U,V}} I(U;Y|V) - I(U;S|V) + \mathsf{R_h} \quad (6.15)$$

$$= \max_{Q_{U|S},Q_{X|U}} I(U;Y) - I(U;S) + \mathsf{R_h}. \quad (6.16)$$

Because the joint distribution is restricted by the Markov condition $S \multimap U \multimap X$, this capacity is, in general, less than the sum of $\mathsf{R_h}$ and the Gel'fand–Pinsker capacity [10], determined by

$$\mathsf{C_{GP}} \triangleq \max_{Q_{U|S},Q_{X|U,S}} I(U;Y) - I(U;S); \quad (6.17)$$

this upper bound would correspond to a scenario where the helper can help the encoder achieve the Gel'fand–Pinsker capacity, while at the same time sending information bits to the receiver at the maximum possible rate $\mathsf{R_h}$.

Choosing $U = (X,S)$ and V deterministic yields the lower bound

$$\mathsf{C}^{(\mathrm{I})}(\mathsf{R_h}) \geq \max_{Q_{X|S}} I(X,S;Y) - H(S) + \mathsf{R_h} \quad (6.18)$$

$$= \max_{Q_{X|S}} \{I(X;Y|S) + I(S;Y)\} + \big(\mathsf{R_h} - H(S)\big) \quad (6.19)$$

$$\geq \max_{Q_{X|S}} I(X;Y|S) + \big(\mathsf{R_h} - H(S)\big) \quad (6.20)$$

with the final inequality possibly strict. Thus, even when $\mathsf{R_h} = H(S)$, the capacity with a message-cognizant helper is in general larger than with a message-oblivious helper because the latter equals $\max_{Q_{X|S}} I(X;Y|S)$ (namely the capacity in the absence of a helper but when the state is known to both encoder and decoder).

Channels that sometimes get stuck. Consider the case where the state alphabet is $\mathcal{S} = \mathcal{Y} \cup \{\mathrm{OK}\}$; the channel behaves like some DMC W

if $S = \mathrm{OK}$, and is stuck at $Y = s$ otherwise:

$$Q_{Y|X,S}(y|x,s) = \begin{cases} W(y|x) & s = \mathrm{OK} \\ \mathbb{1}\{y = s\} & s \in \mathcal{Y}; \end{cases} \tag{6.21}$$

and the state distribution, when conditional on $S \neq \mathrm{OK}$, equals the capacity achieving output distribution of W, denoted $Q_Y^\star$:

$$Q_S(y) = \rho\, Q_Y^\star(y), \qquad y \in \mathcal{Y}, \tag{6.22}$$

where $\rho = 1 - Q_S(\mathrm{OK})$. One special case of the above channel model is a memory with stuck-at faults [57], [58, Example 7.3].

On this channel, it is known that the Gel'fand–Pinsker capacity (6.17) can be maximized by choosing $X = U$ [26, Section 18.6]. Therefore, whenever $\mathsf{R}_\mathrm{h} \geq H(S)$, by choosing $X = U$ also in (6.16), we can indeed achieve

$$\mathsf{C}^{(\mathrm{I})}(\mathsf{R}_\mathrm{h}) = \max_{Q_{U|S},Q_{X|U}} I(U;Y) - I(U;S) + \mathsf{R}_\mathrm{h} \tag{6.23}$$

$$= \max_{Q_{X|S}} I(X;Y) - I(X;S) + \mathsf{R}_\mathrm{h} \tag{6.24}$$

$$- \mathsf{C}_{\mathrm{GP}} + \mathsf{R}_\mathrm{h}. \tag{6.25}$$

Indeed, the relation (6.25) holds also on general SD-DMCs whenever C_{GP} is achieved by choosing $X = U$.

MMANC. Consider the case where $\mathcal{X} = \mathcal{Y} = \mathcal{S} = \mathcal{A} = \{0, 1, \ldots, |\mathcal{A}| - 1\}$ and, with probability one,

$$Y = X \oplus S, \tag{6.26}$$

where "$\oplus$" denotes the mod-$|\mathcal{A}|$ addition. Using (6.7), we next show that in this case

$$\mathsf{C}^{(\mathrm{I})}(\mathsf{R}_\mathrm{h}) = \log|\mathcal{A}| - H(S) + \mathsf{R}_\mathrm{h}. \tag{6.27}$$

We first derive an upper bound:

$$\mathsf{C}^{(\mathrm{I})}(\mathsf{R}_\mathrm{h}) = \max_{\substack{Q_V,Q_{U|S,V},Q_{X|U,V}: \\ I(U;S|V) \leq \mathsf{R}_\mathrm{h}}} I(U;Y|V) - I(U;S|V) + \mathsf{R}_\mathrm{h} \tag{6.28}$$

$$= \max_{\substack{Q_V,Q_{U|S,V},Q_{X|U,V}: \\ I(U;S|V) \leq \mathsf{R}_\mathrm{h}}} I(U;X \oplus S|V) - I(U;S|V) + \mathsf{R}_\mathrm{h} \tag{6.29}$$

$$= \max_{\substack{Q_V, Q_{U|S,V}, Q_{X|U,V}: \\ I(U;S|V) \leq \mathsf{R_h}}} H(X \oplus S|V) - H(X \oplus S|U,V)$$
$$- H(S) + H(S|U,V) + \mathsf{R_h} \quad (6.30)$$

$$\leq \log |\mathcal{A}| - H(S) + \mathsf{R_h}$$
$$+ \max_{\substack{Q_V, Q_{U|S,V}, Q_{X|U,V}: \\ I(U;S|V) \leq \mathsf{R_h}}} H(S|U,V) - H(X \oplus S|U,V) \quad (6.31)$$

$$\leq \log |\mathcal{A}| - H(S) + \mathsf{R_h}, \quad (6.32)$$

where (6.30) holds because $V \perp\!\!\!\perp S$ (indicating that V and S are independent); and (6.32) because $X \multimap (U,V) \multimap S$, and because adding an independent random variable to S increases its entropy.

To obtain a lower bound, consider choosing V deterministic and $U = X = B$, where B is independent of S and equiprobably distributed on $\mathcal{A}$, i.e., $B \perp\!\!\!\perp S$ and $B \sim \mathrm{Unif}(\mathcal{A})$. This yields

$$\mathsf{C}^{(\mathrm{I})}(\mathsf{R_h}) \geq I(U;Y) - I(U;S) + \mathsf{R_h} \quad (6.33)$$
$$= I(B; B \oplus S) - I(B;S) + \mathsf{R_h} \quad (6.34)$$
$$= H(B \oplus S) - H(B \oplus S|B) + \mathsf{R_h} \quad (6.35)$$
$$= \log |\mathcal{A}| - H(S) + \mathsf{R_h}, \quad (6.36)$$

where, in the last step, we used the fact that $B \oplus S$ is uniform over $\mathcal{A}$ (because B is uniform), and that $H(B \oplus S|B) = H(S|B) = H(S)$.

In fact, this capacity can also be achieved with the simple scheme in which the helper ignores the states and uses the entire rate $\mathsf{R_h}$ to send information bits to the receiver.

6.3   Gaussian Channel: Main Results

6.3.1   Problem Formulation

In the Gaussian setting, the time-k channel output Y_k corresponding to the time-k input x_k is

$$Y_k = x_k + Z_k \quad (6.37)$$

where $\{Z_k\}$ are IID variance-N centered Gaussians, i.e., $\sim$ IID $\mathcal{N}(0, \mathsf{N})$. We assume that $\mathsf{N} > 0$, so noise is present. We view the noise as state, so Z_k corresponds to the time-k channel state. With this view in mind, the definition of an achievable rate is very similar to the one for an SD-DMC.

The main difference is that we introduce an average-power constraint on the transmitted signal; see (6.40) ahead.

A blocklength-n rate-R coding scheme with rate-R_h message-cognizant assistance comprises a message set $\mathcal{M} = \{1, \ldots, 2^{nR}\}$; a description set $\mathcal{T} = \{1, \ldots, 2^{nR_h}\}$; a helper that is represented by the Borel measurable helping function

$$h\colon \mathbb{R}^n \times \mathcal{M} \to \mathcal{T} \tag{6.38a}$$

$$(z^n, m) \mapsto t; \tag{6.38b}$$

an encoding function

$$f\colon \mathcal{M} \times \mathcal{T} \to \mathbb{R}^n \tag{6.39a}$$

$$(m, t) \mapsto x^n; \tag{6.39b}$$

that produces the n-tuple $X^n = (X_1(M, T), \ldots, X_n(M, T))$ satisfying,

$$\frac{1}{n}\mathbb{E}_T\left[\sum_{k=1}^{n} X_k^2(m, T)\right] \leq P, \qquad m \in \mathcal{M}, \tag{6.40}$$

where $\mathbb{E}_T[\cdot]$ denotes expectation with respect to T, and where

$$P > 0 \tag{6.41}$$

is some prespecified positive constant; and a decoder that produces an estimate of the message following some decoding rule

$$\phi\colon \mathbb{R}^n \times \mathcal{T} \to \mathcal{M} \tag{6.42a}$$

$$(y^n, t) \mapsto \hat{m}; \tag{6.42b}$$

A rate R is said to be achievable if there exists a sequence of coding schemes as above (indexed by the blocklength) for which

$$\lim_{n\to\infty} \mathbb{P}[M \neq \hat{M}] = 0 \tag{6.43}$$

when M is drawn equiprobably from $\mathcal{M}$. The supremum of achievable rates is the capacity $C(A, R_h)$, where A is the SNR

$$A = \frac{P}{N}. \tag{6.44}$$

The feedback capacity is defined in an analogous way with the transmitted n-tuple $x^n(m,t)$ now having the form

$$x^n(m,t,y^{n-1}) = \big(x_1(m,t), x_2(m,t,y_1), \ldots, x_n(m,t,y^{n-1})\big)$$

and satisfying the second-moment constraint.

The message-oblivious helper capacity with feedback corresponds to a scenario where there is a feedback link as above, but the helper is message-oblivious. The help now has the form $t = \mathsf{h}(z^n)$ and the time-k channel input has the form $x_k(m,t,y^{k-1})$.

6.3.2 Main Results

Theorem 6.7 (Gaussian Channel: Message-Cognizant Helper).
On the Gaussian channel with a noncausal message-cognizant helper that assists both the encoder and the decoder

$$\mathsf{C}(\mathsf{A},\mathsf{R_h}) = \frac{1}{2}\log\left(1 + \mathsf{A} + 2\sqrt{\mathsf{A}(1 - 2^{-2\mathsf{R_h}})}\right) + \mathsf{R_h}. \tag{6.45}$$

This remains the capacity also when a feedback link from the receiver to the encoder is added.

Proof. The proof of the direct part, which does not utilize the feedback link, can be found in Section 6.5.1. The converse, which is valid also in the presence of a feedback link, can be found in Section 6.5.2. $\qquad\square$

As Theorem 6.7 shows, feedback does not increase the capacity of the Gaussian message-cognizant helper capacity (when the help is provided to both the encoder and the decoder). It is, however, useful when the helper is message-oblivious. In the absence of feedback, as in Theorems 2.4, the message-oblivious helper capacity is

$$\frac{1}{2}\log(1 + \mathsf{A}) + \mathsf{R_h}. \tag{6.46}$$

But, as the following theorem shows, feedback increases the capacity to that of message-cognizant helper:

Theorem 6.8 (Gaussian Channel: Message-Oblivious Helper with Feedback). *The capacity of the Gaussian channel with a feedback* *link from the channel output to the encoder and with a noncausal message-oblivious helper that assists both the encoder and the decoder is also*

$$\frac{1}{2} \log\left(1 + \mathsf{A} + 2\sqrt{\mathsf{A}(1 - 2^{-2\mathsf{R_h}})}\right) + \mathsf{R_h} \qquad (6.47)$$

i.e., the same as that with a message-cognizant helper.

Proof. In view of Theorem 6.7, we only need a direct part. This is provided in Section 6.5.3, where we describe a feedback coding scheme with help that does not depend on the message. $\square$

Remark 6.9. As a sanity check, when $\mathsf{R_h} = 0$ (corresponding to the absence of help), the RHS of (6.45) becomes $\frac{1}{2}\log(1 + \mathsf{A})$, as expected. $\triangle$

Remark 6.10. When $\mathsf{R_h} > 0$, both Theorem 6.7 and Theorem 6.8 continue to hold if we replace "channel capacity" with "cutoff rate" and "listsize capacity"; see Section 6.6 for details. $\triangle$

The solid lines in Fig. 6.1a depict the capacity expressions of Theorems 6.7 and 6.8 as a function of $\mathsf{R_h}$ for different values of the SNR A. The dashed lines in that figure correspond to the capacity for a message-oblivious helper in the absence of feedback.

Remark 6.11. In the absence of feedback, the gap between the capacity with a message-oblivious helper (namely (6.46)) and the capacity with a message-cognizant helper (namely the expressions in Theorems 6.7 and 6.8) can be expressed as

$$\left(\frac{1}{2} \log\left(1 + \mathsf{A} + 2\sqrt{\mathsf{A}(1 - 2^{-2\mathsf{R_h}})}\right) + \mathsf{R_h}\right) - \left(\frac{1}{2} \log(1 + \mathsf{A}) + \mathsf{R_h}\right)$$

$$= \frac{1}{2} \log\left(1 + \frac{2\sqrt{1 - 2^{-2\mathsf{R_h}}}}{\sqrt{\mathsf{A}^{-1} + \mathsf{A} + 2}}\right), \qquad (6.48)$$

which depends on the SNR A only via $\mathsf{A} + \mathsf{A}^{-1}$. As a function of the SNR, when the latter is expressed in decibels [dB], this gap is thus a symmetric function. It attains its maximum at $\mathsf{A} = 0$ dB. The solid lines in Fig. 6.1b depict this gap as a function of A [dB] for different values of $\mathsf{R_h}$. $\triangle$

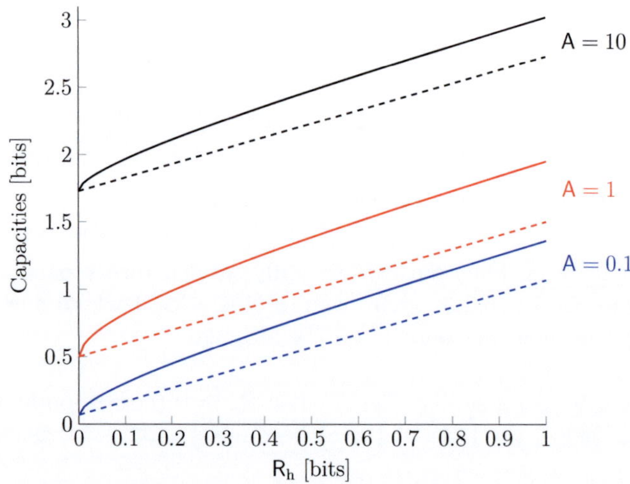

(a) Comparison between the capacity result of Theorems 6.7 and 6.8 (Eq. (6.45), represented by the solid lines) and the message-oblivious helper capacity in the absence of feedback (Eq. (6.46), represented by the dashed lines) as a function of the assistance rate R_h for different values of the SNR A.

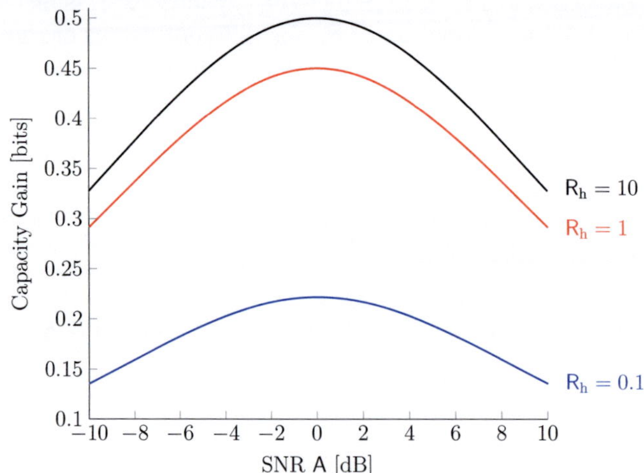

(b) The capacity gains afforded by the message cognition or the feedback link (6.48) as a function of the SNR A [dB] for different values of the assistance rate R_h [bits]

Figure 6.1: Illustrations of the capacity result of Theorems 6.7 and 6.8

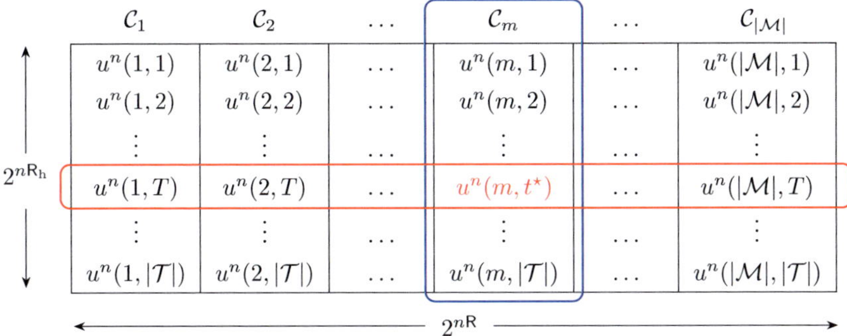

Figure 6.2: Coding scheme with message-cognizant bi-terminal help

6.4 SD-DMC: Proofs

6.4.1 Achievability

We first show how to achieve any rate below

$$\max_{\substack{Q_{U|S}, Q_{X|U}:\\ I(U;S)\leq R_h}} I(U;Y). \tag{6.49}$$

This and a time-sharing argument will then imply the achievability of any rate below $C^{(I)}(R_h)$. By the convexity of $I(U;Y)$ in the conditional distribution of Y given U, the maximum in (6.49) is unaltered when we restrict $Q_{X|U}$ to be deterministic. It thus suffices to prove the achievability of $I(U;Y)$ when $Q_{U|S}$ is such that $I(U;S) < R_h$ and for $X = \psi(U)$ for some arbitrary mapping $\psi\colon \mathcal{U} \to \mathcal{X}$.

Generate 2^{nR} length-n codebooks independently, denoted $\mathcal{C}_m = \{u^n(m,t)\}_{t\in\mathcal{T}}$ for $m \in \mathcal{M}$, each of size 2^{nR_h}, with the n components of each codeword being drawn IID according to the marginal distribution Q_U. This leads to a table of codeword as is illustrated in Fig. 6.2, where the m-th column corresponds to $\mathcal{C}_m$.

If the message to be transmitted is $m \in \mathcal{M}$, and if it observes the channel state sequence s^n, the helper searches the 2^{nR_h} codewords in $\mathcal{C}_m$ (i.e., the blue box in Fig. 6.2) for some $u^n(m,t^\star)$ that is strongly jointly typical with s^n with respect to the (U,S)-marginal $Q_{U,S}$. The helper is very likely to find such a $t^\star$ because $I(U;S) < R_h$. Having found $t^\star$, the helper reveals it to the encoder and the decoder. The encoder now transmits $\psi(u^n(m,t^\star))$, where $\psi(\cdot)$ extends to n-tuples componentwise. The decoder, with $t^\star$ and the received sequence y^n in hand, searches among the codewords in the red box in Fig. 6.2 for some $\hat{m}$, for which $u^n(\hat{m}, t^\star)$ is strongly jointly typical with y^n with respect to the (U, Y)-marginal $Q_{U,Y}$.

We next analyze the scheme. Whenever the helper succeeds in finding a suitable $t^\star$ (which, we recall, happens with high probability), the transmitted codeword u^n is strongly jointly typical with s^n. In this case, since x^n is a deterministic function of u^n, the triple (u^n, s^n, x^n) must also be strongly jointly typical. Since $U \mathrel{-\!\circ\!-} (X, S) \mathrel{-\!\circ\!-} Y$ forms a Markov chain under the joint PMF at hand, and since Y^n is the channel response to (x^n, s^n), it follows from the Markov Lemma [58, Lemma 12.1], [26, Corollary 4.12] that, with high probability, Y^n is strongly jointly typical with u^n.

As to an incorrect codeword, say $u^n(m', t^\star)$ for some $m' \neq m$, it is drawn independently of Y^n, so the probability that it is jointly typical with Y^n is approximately $2^{-nI(U;Y)}$. Since there are $(2^{nR} - 1)$ such codewords, the average error probability will vanish with the blocklength whenever $\mathsf{R} < I(U;Y)$. This establishes the achievability of all rates smaller than (6.49).

Next, we introduce a time-sharing random variable V. For each of its realizations, say $V = v$, we construct a coding scheme as above with transmission rate R_v and assistance rate $\mathsf{R}_{\mathrm{h},v}$, to be used in $n \cdot Q_V(v)$ of the n channel uses in a block. This allows us to achieve all rates below

$$\max_{\substack{Q_V, Q_{U|S,V}, Q_{X|U,V}: \\ I(U;S|V) \leq \mathsf{R}_\mathrm{h}}} I(U;Y|V). \tag{6.50}$$

Finally, we introduce R_0, which corresponds to the rate at which the helper sends information bits directly to the receiver. Specifically, the assistance T is a tuple $(T_1, \tilde{M})$, where T_1 is of rate $\mathsf{R}_\mathrm{h} - \mathsf{R}_0$, which is substituted for the help rate in the scheme above (with the joint distribution now satisfying $I(U;S|V) \leq \mathsf{R}_\mathrm{h} - \mathsf{R}_0$), while $\tilde{M}$ is of rate R_0, and is part of the message to be communicated. The total transmission rate is then $I(U;Y|V) + \mathsf{R}_0$. Optimizing over R_0 and the distributions yields the achievability of the RHS of (6.9) and establishes the direct part of Theorem 6.5. $\qquad\qquad\square$

6.4.2 Converse

We now prove a converse in the presence of a feedback link from the channel output to the encoder.

Consider a message M that is drawn equiprobably from $\mathcal{M}$. Fano's inequality implies that, for any sequence of rate-R coding schemes with rate-R_h message-cognizant assistance and vanishing average probability of error, there exists some sequence $\{\delta_n\}$ tending to zero such that

$$n R - n\delta_n = H(M) - H(M|Y^n, T) \tag{6.51}$$
$$= I(M; Y^n, T) \tag{6.52}$$
$$= I(M; Y^n|T) + I(M; T) \tag{6.53}$$
$$= H(Y^n|T) - H(Y^n|M, T) + I(M; T) \tag{6.54}$$
$$= \sum_{k=1}^{n} H(Y_k|T, Y^{k-1}) - H(Y_k|M, T, Y^{k-1}) + I(M; T) \tag{6.55}$$
$$\leq \sum_{k=1}^{n} H(Y_k) - H(Y_k|M, T, Y^{k-1}) + I(M; T) \tag{6.56}$$
$$= \sum_{k=1}^{n} I(Y_k; U_k) + I(M; T), \tag{6.57}$$
$$= n\, I(Y_V; U_V|V) + n\bar{R}_0, \tag{6.58}$$

where in (6.57) we introduce $U_k \triangleq (M, T, Y^{k-1})$ for $k \in [n]$; and in (6.58) we define $\bar{R}_0 \triangleq \frac{1}{n} I(M; T)$, and introduce an auxiliary random variable V that is equiprobable over $\{1, \ldots, n\}$ and is independent of all other random quantities. Dividing by n, we have

$$R - \delta_n \leq I(Y_V; U_V|V) + \bar{R}_0. \tag{6.59}$$

To relate this upper bound to (6.9), we proceed to justify that, by our choice of V and U_V, the joint distribution of (V, U_V, X_V, S_V, Y_V) satisfies the conditions in the maximization in (6.9).

First, we verify that the joint distribution factorizes correctly as

$$Q_V \circ Q_{S_V} \circ Q_{U_V|S_V, V} \circ Q_{X_V|U_V, V} \circ Q_{Y_V|X_V, S_V}. \tag{6.60}$$

Because $\{S_k\} \sim \text{IID } Q_S$, we have $S_V \perp\!\!\!\perp V$. The Markov relation $S_V \multimap (U_V, V) \multimap X_V$ holds because X_V, as a function of (M, T, Y^{V-1}, V), is a function of (U_V, V). Finally, the Markov relation $(V, U_V) \multimap (X_V, S_V) \multimap Y_V$ holds because the channel is memoryless.

We next show that $I(S_V; U_V | V) \leq \mathsf{R_h} - \bar{\mathsf{R}}_0$ as follows:.

$$n\, I(S_V; U_V | V) = \sum_{k=1}^{n} I(S_k; U_k) \tag{6.61}$$

$$= \sum_{k=1}^{n} I(S_k; M, T, Y^{k-1}) \tag{6.62}$$

$$\leq \sum_{k=1}^{n} I(S_k; M, T, Y^{k-1}, S^{k-1}) \tag{6.63}$$

$$= \sum_{k=1}^{n} I(S_k; M, T, S^{k-1}) \tag{6.64}$$

$$= \sum_{k=1}^{n} I(S_k; M, T | S^{k-1}) \tag{6.65}$$

$$= I(S^n; M, T) \tag{6.66}$$

$$= I(S^n; T | M) \tag{6.67}$$

$$= I(S^n, M; T) - I(M; T) \tag{6.68}$$

$$\leq \log |\mathcal{T}| - I(M; T) \tag{6.69}$$

$$= n\mathsf{R_h} - n\bar{\mathsf{R}}_0, \tag{6.70}$$

where (6.64) holds because, given (M, T, S^{k-1}), the output Y^{k-1} is a function only of the channel randomness and is thus independent of all other random variables.

Finally, letting n tend to infinity yields the desired converse. $\qquad\square$

6.5   Gaussian Channel: Proofs

6.5.1   Achievability

6.5.1.1   In Broad Brushstrokes

We begin with a rough description of the coding scheme that ignores some of the technicalities. This captures the main idea of the scheme and intuitively explains where the gain comes from. Let F_{XYZ} be the centered multivariate Gaussian distribution under which (X, Z) are of covariance matrix

$$\begin{pmatrix} \mathsf{P} & \sqrt{\mathsf{PN}}\rho \\ \sqrt{\mathsf{PN}}\rho & \mathsf{N} \end{pmatrix} \tag{6.71}$$

and

$$Y = X + Z \tag{6.72}$$

with probability one, where

$$\rho = \sqrt{1 - 2^{-2\mathsf{R_h}}} \tag{6.73}$$

so

$$I(X;Z) = h(Z) - h(Z|X) \tag{6.74}$$

$$= \frac{1}{2}\log(2\pi e\mathsf{N}) - \frac{1}{2}\log(2\pi e\mathsf{N}(1-\rho^2)) \tag{6.75}$$

$$= \mathsf{R_h}. \tag{6.76}$$

Generate $2^{n(\mathsf{R}+\mathsf{R_h})}$ codewords $\{x^n(m,t)\}_{(m,t)\in\mathcal{M}\times\mathcal{T}}$ independently, with the n components of each being drawn IID $\mathcal{N}(0,\mathsf{P})$. If the message to be transmitted is m, and if it observes the noise sequence Z^n, the helper searches the $2^{n\mathsf{R_h}}$ codewords $\{x^n(m,t)\}_{t\in\mathcal{T}}$ for a codeword $x^n(m,t^\star)$ that is (weakly) jointly typical with Z^n with respect to the XZ-marginal F_{XZ} of the above distribution F_{XYZ}. The helper is very likely to find such $t^\star$ because, by our choice of ρ (6.73), $\mathsf{R_h} \approx I(X;Z)$. Having found $t^\star$, the helper reveals it to the encoder and the decoder, with the former now transmitting $X^n = x^n(m,t^\star)$. The decoder, for its part, searches $\{x^n(m',t^\star)\}_{m'\in\mathcal{M}}$ for a some $\hat{m}$ for which $x^n(\hat{m},t^\star)$ is jointly typical with the received sequence Y^n with respect to the XY-marginal F_{XY} of the above F_{XYZ}.

It can be verified that if X^n and Z^n are jointly weakly typical with respect to F_{XZ}, then X^n and $Y^n = X^n + Z^n$ are also jointly typical with respect to F_{XY}. Since the incorrect codewords are drawn independently of Y^n, the decoding will succeed with high probability when R is approximately $I(X;Y)$ (where the latter is computed with respect to F_{XY}). This mutual information is given by

$$I(X;Y) = h(Y) - h(Y|X) \tag{6.77}$$

$$= h(Y) - h(Z|X) \tag{6.78}$$

$$= \frac{1}{2}\log\left(2\pi e(\mathsf{P} + \mathsf{N} + 2\sqrt{\mathsf{PN}}\rho)\right) - \frac{1}{2}\log(2\pi e\mathsf{N}(1-\rho^2)) \tag{6.79}$$

$$= \frac{1}{2}\log\left(1 + \mathsf{A} + 2\sqrt{\mathsf{A}(1 - 2^{-2\mathsf{R_h}})}\right) + \mathsf{R_h}. \tag{6.80}$$

6.5.1.2   A Geometric Approach

For a more rigorous achievability proof, we propose a geometric approach.

Let $\partial\mathcal{B}(\sqrt{n\mathsf{P}}) = \{x^n \in \mathbb{R}^n : \|x^n\|^2 = n\mathsf{P}\}$ denote the radius-$\sqrt{n\mathsf{P}}$ $(n-1)$-dimensional Euclidean sphere in $\mathbb{R}^n$, and let $\angle(x^n, y^n) \in [0, \pi]$ denote the angle between the two (nonzero) vectors $x^n, y^n \in \mathbb{R}^n$ in the sense that

$$\cos \angle(x^n, y^n) = \frac{\langle x^n, y^n \rangle}{\|x^n\|\|y^n\|}. \tag{6.81}$$

Fix $\epsilon \in (0, \mathsf{R_h})$ (later to tend to zero), and let $\theta_0 \in [0, \pi/2]$ be such that

$$\sin \theta_0 = 2^{-(\mathsf{R_h} - \epsilon)}. \tag{6.82}$$

Let $\mathcal{C} \subset \partial\mathcal{B}(\sqrt{n\mathsf{P}})$ be a codebook of $2^{n\mathsf{R_h}}$ codewords, indexed by $\mathcal{T}$, with the covering property that the caps of half-angle θ_0 centered around the codewords completely cover $\partial\mathcal{B}(\sqrt{n\mathsf{P}})$. Such a codebook exists whenever n is large enough [59], as we henceforth assume.

Pick $|\mathcal{M}|$ random orthogonal transformations (rotations) independently, each uniform according to the Haar measure, and index them by the messages $m \in \mathcal{M}$. For each $m \in \mathcal{M}$, generate the set $\mathcal{C}_m = \{x^n(m, t)\}_{t \in \mathcal{T}}$ by applying the orthogonal transformation corresponding to m to each of the codewords in $\mathcal{C}$.

Note that for each $m \in \mathcal{M}$, the set $\mathcal{C}_m$—being the result of rotating $\mathcal{C}$—also satisfies the covering property. This will be important to keep in mind when we describe the transmission scheme. Also note that, for each fixed $t \in \mathcal{T}$, the codewords $\{x^n(m, t)\}_{m \in \mathcal{M}}$—which are the result of applying different random rotations to the same element of $\mathcal{C}$—are independent and uniformly distributed over the sphere. This observation will be crucial to our analysis of the probability of error.

We next describe the transmission of some $m \in \mathcal{M}$. Upon observing the noise Z^n, the helper seeks some $t^\star \in \mathcal{T}$ such that the angle between $x^n(m, t^\star)$ and Z^n does not exceed θ_0. Such a $t^\star$ exists because $\mathcal{C}_m$ inherits the covering property from $\mathcal{C}$. This $t^\star$ (or one of those satisfying the requirement) is revealed to both the encoder and the decoder, with the former now transmitting $x^n(m, t^\star)$.

The decoder—based on its observation Y^n and the help $t^\star$—produces

$$\hat{m} = \arg \min_{m' \in \mathcal{M}} \|Y^n - x^n(m', t^\star)\|. \tag{6.83}$$

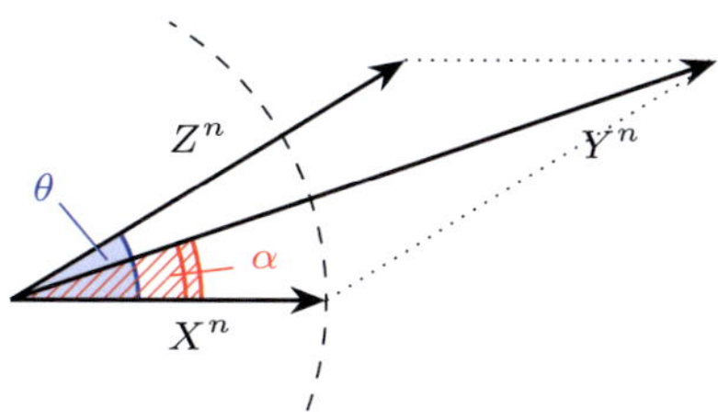

Figure 6.3: Geometric illustration of the sequences

We next analyze the probability of error of our scheme. To that end, we illustrate the geometric relation in Fig. 6.3. Set

$$\theta = \angle(X^n, Z^n) \tag{6.84}$$
$$\leq \theta_0 \tag{6.85}$$

where the inequality follows from our choice of $t^\star$. In terms of θ,

$$\|Y^n\|^2 = \|X^n\|^2 + \|Z^n\|^2 + 2\|X^n\|\|Z^n\|\cos\theta. \tag{6.86}$$

Setting

$$\alpha = \angle(X^n, Y^n) \tag{6.87}$$

we observe that

$$\sin\alpha = \frac{\|Z^n\|}{\|Y^n\|}\sin\theta \tag{6.88}$$

$$= \left(\sqrt{\frac{\|X^n\|^2}{\|Z^n\|^2} + 1 + 2\frac{\|X^n\|}{\|Z^n\|}\cos\theta}\right)^{-1}\sin\theta. \tag{6.89}$$

Recalling that $\|X^n\| = \sqrt{n\mathsf{P}}$, we obtain that, whenever $\|Z^n\|^2 \leq n(\mathsf{N}+\epsilon)$,

$$\sin\alpha \leq \left(\sqrt{\frac{\mathsf{P}}{\mathsf{N}+\epsilon} + 1 + 2\sqrt{\frac{\mathsf{P}}{\mathsf{N}+\epsilon}}\cos\theta}\right)^{-1}\sin\theta \tag{6.90}$$

$$\leq \left(\sqrt{\frac{\mathsf{P}}{\mathsf{N}+\epsilon} + 1 + 2\sqrt{\frac{\mathsf{P}}{\mathsf{N}+\epsilon}}\cos\theta_0}\right)^{-1}\sin\theta_0 \tag{6.91}$$

$$\triangleq \sin\alpha_0, \tag{6.92}$$

where (6.91) holds because $\theta \leq \theta_0$ and (6.92) defines $\alpha_0 \in [0, \pi/2]$.

Having verified that the condition $\|Z^n\|^2 \leq n(\mathsf{N} + \epsilon)$ implies that $\angle(X^n, Y^n) \leq \alpha_0$ and that the condition $m' \neq m$ implies that $x^n(m', T)$ is independent of Y^n and uniformly distributed over the sphere, we can bound the probability of error as follows:

$$P_e(m) \leq \mathbb{P}\big[\|Z^n\|^2 > n(\mathsf{N} + \epsilon)\big]$$
$$+ \mathbb{P}\big[\exists m' \neq m : \angle(x^n(m', T), Y^n) \leq \alpha_0\big] \quad (6.93)$$
$$\leq \mathbb{P}\big[\|Z^n\|^2 > n(\mathsf{N} + \epsilon)\big] + 2^{n\mathsf{R}} \cdot \frac{C_n(\alpha_0)}{C_n(\pi)} \quad (6.94)$$
$$= \mathbb{P}\big[\|Z^n\|^2 > n(\mathsf{N} + \epsilon)\big] + 2^{n\mathsf{R}} \cdot 2^{-n(\log \sin \alpha_0 + o(1))}, \quad (6.95)$$

where in (6.94) we use $C_n(\gamma)$ to denote the surface area of a spherical cap of half-angle γ on a unit n-sphere for $\gamma \in [0, \pi]$; and (6.95) follows from [60] (see also [59]). The upper bound (6.95) decays to zero whenever

$$\mathsf{R} < -\log \sin \alpha_0 \quad (6.96)$$

$$= -\log\left(\left(\sqrt{\frac{\mathsf{P}}{\mathsf{N}+\epsilon} + 1 + 2\sqrt{\frac{\mathsf{P}}{\mathsf{N}+\epsilon}} \cos \theta_0}\right)^{-1} \sin \theta_0\right) \quad (6.97)$$

$$= \frac{1}{2} \log\left(\frac{\mathsf{P}}{\mathsf{N}+\epsilon} + 1 + 2\sqrt{\frac{\mathsf{P}}{\mathsf{N}+\epsilon}} \cos \theta_0\right) - \log \sin \theta_0 \quad (6.98)$$

$$= \frac{1}{2} \log\left(\frac{\mathsf{P}}{\mathsf{N}+\epsilon} + 1 + 2\sqrt{\frac{\mathsf{P}}{\mathsf{N}+\epsilon}} \sqrt{1 - 2^{-2(\mathsf{R_h}-\epsilon)}}\right) + \mathsf{R_h} - \epsilon. \quad (6.99)$$

The direct part is now concluded by letting ϵ tend to zero and by employing the random-coding argument that guarantees that there exists a deterministic unitary transformation resulting in an arbitrarily small probability of error. $\qquad\square$

6.5.2   Converse

We now prove a converse in the presence of a feedback link from the channel output to the encoder. Consider a message M that is drawn equiprobably from $\mathcal{M}$. Fano's inequality implies that, for any sequence of rate-R coding schemes with rate-$\mathsf{R_h}$ message-cognizant assistance and vanishing probabilities of error, there exists some sequence $\{\delta_n\}$ tending to zero such that

$$nR - n\delta_n = H(M) - H(M|Y^n, T) \tag{6.100}$$
$$= I(M; Y^n, T) \tag{6.101}$$
$$= I(M; Y^n|T) + I(M; T) \tag{6.102}$$
$$= h(Y^n|T) - h(Y^n|M, T) + I(M; T) \tag{6.103}$$
$$= h(Y^n|T) - h(Z^n) + I(Z^n; T|M) + I(M; T) \tag{6.104}$$
$$= h(Y^n|T) - h(Z^n) + I(Z^n, M; T) \tag{6.105}$$
$$\leq h(Y^n|T) - h(Z^n) + \log |\mathcal{T}| \tag{6.106}$$
$$\leq \sum_{k=1}^{n} h(Y_k) - h(Z^n) + \log |\mathcal{T}|, \tag{6.107}$$

where (6.107) holds because conditioning reduces entropy, and (6.104) can be justified as follows:

$$h(Y^n|M, T) = \sum_{k=1}^{n} h(Y_k|M, T, Y^{k-1}) \tag{6.108}$$
$$= \sum_{k=1}^{n} h(Y_k - X_k|M, T, Y^{k-1}) \tag{6.109}$$
$$= \sum_{k=1}^{n} h(Z_k|M, T, Y^{k-1}) \tag{6.110}$$
$$= \sum_{k=1}^{n} h(Z_k) - \sum_{k=1}^{n} I(Z_k; M, T, Y^{k-1}) \tag{6.111}$$
$$= \sum_{k=1}^{n} h(Z_k) - \sum_{k=1}^{n} I(Z_k; M, T, Z^{k-1}) \tag{6.112}$$
$$= h(Z^n) - \sum_{k=1}^{n} I(Z_k; M, T, Z^{k-1}) \tag{6.113}$$
$$= h(Z^n) - I(Z^n; T|M) \tag{6.114}$$

where (6.109) holds because X_k is a function of (M, T, Y^{k-1}); (6.112) holds because there is a bijection between (M, T, Y^{k-1}) and (M, T, Z^{k-1}); and (6.114) holds because

$$\sum_{k=1}^{n} I(Z_k; M, T, Z^{k-1}) = \sum_{k=1}^{n} I(Z_k; M, T|Z^{k-1}) \tag{6.115}$$

$$= I(Z^n; M, T) \tag{6.116}$$

$$= I(Z^n; T|M). \tag{6.117}$$

Having justified (6.107), it remains to upper-bound its RHS. We begin by bounding $I(X_k; Z_k)$ in two different ways. The first upper-bounds it:

$$\sum_{k=1}^{n} I(X_k; Z_k) \leq \sum_{k=1}^{n} I(X_k, M, T, Z^{k-1}; Z_k) \tag{6.118}$$

$$= \sum_{k=1}^{n} I(M, T, Z^{k-1}; Z_k) \tag{6.119}$$

$$= I(Z^n; T|M) \tag{6.120}$$

$$\leq \log |\mathcal{T}| \tag{6.121}$$

$$= n\mathsf{R_h}, \tag{6.122}$$

where (6.119) holds because X_k is a function of (M, T, Z^{k-1}); and (6.120) follows from (6.117). The second lower-bounds it:

$$I(X_k; Z_k) = h(Z_k) - h(Z_k|X_k) \tag{6.123}$$

$$= \frac{1}{2} \log(2\pi e \mathsf{N}) - h(Z_k|X_k) \tag{6.124}$$

$$\geq \frac{1}{2} \log(2\pi e \mathsf{N}) - \frac{1}{2} \log(2\pi e \mathsf{N}(1 - \rho_k^2)) \tag{6.125}$$

$$= -\frac{1}{2} \log(1 - \rho_k^2), \tag{6.126}$$

where in (6.125) we define $\rho_k \in [-1, 1]$ to be the correlation coefficient between X_k and Z_k, and the inequality follows from [58, Problem 2.7].

From the two bounds (6.122) and (6.126),

$$\mathsf{R_h} \geq \frac{1}{n} \sum_{k=1}^{n} -\frac{1}{2} \log(1 - \rho_k^2) \tag{6.127}$$

$$\geq -\frac{1}{2} \log \left(1 - \frac{1}{n} \sum_{k=1}^{n} \rho_k^2 \right) \tag{6.128}$$

where (6.128) follows from Jensen's inequality and the concavity of the logarithmic function. Hence,

$$\sum_{k=1}^{n} \rho_k^2 \leq n\left(1 - 2^{-2\mathsf{R_h}}\right). \tag{6.129}$$

We now use this inequality to upper-bound the sum on the RHS of (6.107). For each k

$$\mathsf{Var}[Y_k] = \mathsf{Var}[X_k + Z_k] \tag{6.130}$$

$$= \mathsf{Var}[X_k] + \mathsf{Var}[Z_k] + 2\sqrt{\mathsf{Var}[X_k]\,\mathsf{Var}[Z_k]}\rho_k, \tag{6.131}$$

so

$$\sum_{k=1}^{n} h(Y_k) \le \sum_{k=1}^{n} \frac{1}{2} \log\left(2\pi e\,\mathsf{Var}[Y_k]\right) \tag{6.132}$$

$$\le \frac{n}{2} \log\left(2\pi e \cdot \frac{1}{n}\sum_{k=1}^{n} \mathsf{Var}[Y_k]\right) \tag{6.133}$$

$$= \frac{n}{2} \log\left(2\pi e \cdot \frac{1}{n}\sum_{k=1}^{n}\left(\mathsf{Var}[X_k] + \mathsf{Var}[Z_k] \right.\right.$$
$$\left.\left. + 2\sqrt{\mathsf{Var}[X_k]\,\mathsf{Var}[Z_k]}\rho_k\right)\right) \tag{6.134}$$

$$\le \frac{n}{2} \log\left(2\pi e \cdot \frac{1}{n}\sum_{k=1}^{n}\left(\mathbb{E}\left[X_k^2\right] + \mathsf{Var}[Z_k] \right.\right.$$
$$\left.\left. + 2\sqrt{\mathbb{E}[X_k^2]\,\mathsf{Var}[Z_k]}\,|\rho_k|\right)\right) \tag{6.135}$$

$$\le \frac{n}{2} \log\left(2\pi e\left(\mathsf{P} + \mathsf{N} + \frac{2}{n}\sum_{k=1}^{n}\sqrt{\mathsf{N}\mathbb{E}[X_k^2]}\,|\rho_k|\right)\right) \tag{6.136}$$

$$\le \frac{n}{2} \log\left(2\pi e\left(\mathsf{P} + \mathsf{N} + \frac{2}{n}\sqrt{\mathsf{N}\sum_{k=1}^{n}\mathbb{E}[X_k^2]}\sqrt{\sum_{k=1}^{n}\rho_k^2}\right)\right) \tag{6.137}$$

$$\le \frac{n}{2} \log\left(2\pi e\left(\mathsf{P} + \mathsf{N} + \frac{2}{n}\sqrt{n\mathsf{P}\mathsf{N}}\sqrt{n\left(1 - 2^{-2\mathsf{R}_\mathrm{h}}\right)}\right)\right) \tag{6.138}$$

$$= \frac{n}{2} \log\left(2\pi e\left(\mathsf{P} + \mathsf{N} + 2\sqrt{\mathsf{P}\mathsf{N}\left(1 - 2^{-2\mathsf{R}_\mathrm{h}}\right)}\right)\right) \tag{6.139}$$

where (6.133) follows from the concavity of the logarithmic function; (6.137) follows from the Cauchy–Schwarz inequality; and (6.138) follows from (6.129).

Continuing from (6.107),

$$n\mathsf{R} - n\delta_n \le \sum_{k=1}^{n} h(Y_k) - \frac{n}{2}\log(2\pi e\mathsf{N}) + n\mathsf{R}_\mathrm{h} \tag{6.140}$$

$$\leq n \cdot \frac{1}{2} \log\left(2\pi e\left(\mathsf{P} + \mathsf{N} + 2\sqrt{\mathsf{PN}\left(1 - 2^{-2\mathsf{R_h}}\right)}\right)\right)$$

$$- \frac{n}{2} \log(2\pi e\mathsf{N}) + n\mathsf{R_h} \qquad (6.141)$$

$$= n \cdot \frac{1}{2} \log\left(1 + \mathsf{A} + 2\sqrt{\mathsf{A}\left(1 - 2^{-2\mathsf{R_h}}\right)}\right) + n\mathsf{R_h}. \qquad (6.142)$$

Dividing both sides of the inequality by n and letting n tend to infinity yields the desired inequality. $\qquad\square$

6.5.3   Feedback Scheme for Message-Oblivious Helper

We next prove Theorem 6.8 by proposing a feedback scheme with a helper that is incognizant of the message. The scheme is reminiscent of the Schalkwijk–Kailath scheme [61, 62], albeit with only one use of the feedback link. For notational reasons, we consider transmission with blocks of length $n + 1$ and denote the corresponding channel inputs $X_0, X_1, \ldots, X_n$ for positive integers n.

In broad brushstrokes, the idea is the following: We map the message m to X_0 using an injective mapping so that from X_0 one could recover the message. The problem is, of course, that the receiver has no access to X_0 but only to Y_0, i.e., to the sum of X_0 and the noise sample Z_0. This noise sample, however, is known to the encoder as of time 1, because it is simply the difference between Y_0 and X_0, and the former is revealed to the encoder after time zero. Moreover, the noise sample Z_0 is known to the helper ahead of time, and *a fortiori* as of time 1. We can then think about Z_0 as a new message that is to be conveyed to the receiver using $X_1, \ldots, X_n$ with this message being known to both helper and encoder. The idea is to then invoke Theorem 6.7 for the transmission of Z_0 in the time slots 1 through n. The noise sample Z_0 is, of course, continuous, so there are some quantization issues to be addressed.

To address these, we use a construction somewhat similar to the one in Section 2.2.3. For notational reasons we now assume that the message set is $\{0, \ldots, 2^{n\mathsf{R}} - 1\}$ (i.e., starting at zero) and assume that $2^{n\mathsf{R}}$ is an integer (in order to be able to write $2^{n\mathsf{R}}$ instead of $\lfloor 2^{n\mathsf{R}} \rfloor$).

To convey some message $m \in \mathcal{M}$, the encoder transmits at time zero the symbol

$$X_0 = X_0(m) = m \cdot \frac{\sqrt{\mathsf{P}}}{2^{n\mathsf{R}}}. \qquad (6.143)$$

The decoder observes $Y_0 = X_0 + Z_0$, and the encoder—thanks to the feedback link—can calculate $Z_0 = Y_0 - X_0$ after time 0. Using the symbols $X_1, \ldots, X_n$, the encoder attempts to convey to the decoder a quantized version of Z_0 or, more precisely, the integer

$$M' = \left\lfloor Z_0 \cdot \frac{2^{n\mathsf{R}}}{\sqrt{\mathsf{P}}} \right\rfloor \quad \mathrm{mod}\ 2^{n\mathsf{R}}, \tag{6.144}$$

which is a function of Z_0 and hence is also known to the helper. It does so while ignoring the feedback link. The feedback link is thus used by our scheme only to convey Y_0 to the encoder before time 1.

Since, as of time 1, the integer $M' \in \mathcal{M}$ is known to both encoder and helper, we can employ a blocklength-n coding scheme with message-cognizant helper as in Theorem 6.7 to send M' using $X_1, \ldots, X_n$ with arbitrarily small probability of error. With the receiver's guess of M' (based on $Y_1, \ldots, Y_n$) denoted $\hat{M}'$, the receiver then guesses that the transmitted message was

$$\hat{M} = \left\lfloor Y_0 \cdot \frac{2^{n\mathsf{R}}}{\sqrt{\mathsf{P}}} - \hat{M}' \right\rfloor \quad \mathrm{mod}\ 2^{n\mathsf{R}}. \tag{6.145}$$

If the decoder recovers M' correctly, i.e., if $\hat{M}' = M'$, then its guess $\hat{M}$ of m is correct, because in this case

$$\hat{M} = \left\lfloor (X_0 + Z_0) \cdot \frac{2^{n\mathsf{R}}}{\sqrt{\mathsf{P}}} - M' \right\rfloor \quad \mathrm{mod}\ 2^{n\mathsf{R}} \tag{6.146}$$

$$= m + \left\lfloor Z_0 \cdot \frac{2^{n\mathsf{R}}}{\sqrt{\mathsf{P}}} \right\rfloor - M' \quad \mathrm{mod}\ 2^{n\mathsf{R}} \tag{6.147}$$

$$= m. \tag{6.148}$$

The probability of error in reconstructing m is thus upper-bounded by the probability of error in reconstructing M', which, by Theorem 6.7, can be made arbitrarily small whenever

$$\mathsf{R} < \frac{1}{2} \log\left(1 + \mathsf{A} + 2\sqrt{\mathsf{A}\left(1 - 2^{-2\mathsf{R_h}}\right)}\right) + \mathsf{R_h}. \tag{6.149}$$

This proves the achievability part of Theorem 6.8. $\qquad\qquad\qquad\square$

6.6 Gaussian Channel: Cutoff Rate and List-size Capacity

Recall that, given $\rho > 0$, the ρ-th order listsize capacity $C_\ell^{(\rho)}$ is the supremum of the rates for which there exist codes such that the ρ-th moment of the cardinality of the zero-error list $\mathcal{L}_0$ convergences to one. The ρ-th order cutoff rate $R_{\text{cutoff}}^{(\rho)}$ is defined similarly, but with the list now replaced by the at-least-as-likely list $\mathcal{L}$ with respect to the transmitted message. (See Chapter 2 for more details.) By definition,

$$C_\ell^{(\rho)}(R_h) \leq R_{\text{cutoff}}^{(\rho)}(R_h) \leq C(R_h). \tag{6.150}$$

We establish that on the Gaussian channel with a bi-terminal message-cognizant helper, both inequalities in (6.150) hold with equality. Moreover, a counterpart to Theorem 6.8 with a message-oblivious helper also holds.

Theorem 6.12 (Cutoff Rate and Listsize Capacity: Message-Cognizant Helper). *Irrespective of $\rho > 0$ and of whether a feedback link from the receiver to the encoder is present, on the Gaussian channel with a noncausal message-cognizant helper that assists both the encoder and the decoder at rate $R_h > 0$*

$$R_{\text{cutoff}}^{(\rho)}(R_h) = C_\ell^{(\rho)}(R_h) \tag{6.151}$$

$$= \frac{1}{2}\log\left(1 + A + 2\sqrt{A(1 - 2^{-2R_h})}\right) + R_h. \tag{6.152}$$

Theorem 6.13 (Cutoff Rate and Listsize Capacity: Message-Oblivious Helper and Feedback). *Irrespective of $\rho > 0$, the ρ-th order cutoff rate and the ρ-th order listsize capacity of the Gaussian channel with feedback and with a noncausal message-oblivious helper that assists both the encoder and the decoder at rate $R_h > 0$ are both given by*

$$\frac{1}{2}\log\left(1 + A + 2\sqrt{A(1 - 2^{-2R_h})}\right) + R_h \tag{6.153}$$

i.e., they both coincide with the capacity when the helper is message-cognizant.

Proof of Theorem 6.13. The proof follows from Theorem 6.12 in the same way that Theorem 6.8 follows from Theorem 6.7, as in Section 6.5.3. The details are omitted. $\qquad\square$

Proof of Theorem 6.12. Here we first explain intuitively why the theorem holds.

The reason why the listsize capacity equals the cutoff rate for $R_h > 0$ is similar to that of Theorem 4.17: A description of the normalized squared Euclidean norm of Z^n enables the decoder to calculate the likelihood function and causes only negligible loss in assistance rate. This eliminates the distinction between the list $\mathcal{L}_0$ and $\mathcal{L}$.

The cutoff rate can be as high as the capacity due to the following. On the one hand, with a description of the Euclidean norm of Z^n, the power of X^n can scale with Z^n. On the other hand, by utilizing the coding scheme in Section 6.5.1.2, X^n aligns with Z^n with probability one. Therefore, the relation in Fig. 6.3 is guaranteed to hold, and the analysis of the list $\mathcal{L}$ becomes equivalent to the analysis of the probability of error with the Union Bound.

Some details pertaining to the quantization are to be addressed. A detailed proof can be found in Appendix D.2. $\qquad\square$

Part III

Appendices

Appendix A

Proofs for Chapter 3

A.1   Proof of Proposition 3.7

Let $\mathcal{D} \subset \mathbb{R}$ be the collection of all the point masses of Q_Z, i.e., $\mathcal{D} \triangleq \{z \in \mathbb{R} : Q_Z(\{z\}) > 0\}$, so $Q_Z(\mathcal{D}) > 0$ and $\mathcal{D}$ is countable. We now construct an infinite set $\mathcal{I}' = \{\xi_1, \xi_2, \dots\}$ that is contained in $\mathcal{I}$ and whose elements satisfy

$$\left(i \neq j\right) \implies \left((\xi_i + \mathcal{D}) \cap (\xi_j + \mathcal{D}) = \emptyset\right). \tag{A.1}$$

We do so by choosing ξ_1 to be an arbitrary element of $\mathcal{I}$ and then proceeding inductively to choose ξ_k to be an arbitrary element of the set

$$\mathcal{I} \setminus \cup_{i=1}^{k-1}\left(\xi_i + (\mathcal{D} - \mathcal{D})\right), \tag{A.2}$$

where

$$\mathcal{D} - \mathcal{D} \triangleq \left\{\eta_1 - \eta_2 : \eta_1, \eta_2 \in \mathcal{D}\right\}. \tag{A.3}$$

(Since $\mathcal{D}$ is countable, so is $\mathcal{D} - \mathcal{D}$ and hence also the union in (A.2), so (A.2) is not empty because $\mathcal{I}$ is uncountable.)

We propose the following coding scheme, where the input symbols take values only in $\mathcal{I}'$. The encoder sends some $x \in \mathcal{I}'$ repeatedly n times. The decoder, for each $k \in [n]$, searches $\mathcal{I}'$ for some $\xi \in \mathcal{I}'$ such that

$$Y_k \in \xi + \mathcal{D}. \tag{A.4}$$

If such a ξ exists, it decodes the input symbol as ξ (or as one of the ξ's if there is more than one); else, when there are none, the decoder declares an erasure.

The analysis of the performance is again two-folded: we will first show that the probability of erasure tends to zero as n tends to infinity, and then we will prove that the probability of undetected error is zero for all n.

In the proposed scheme, an erasure occurs only if none of the noise symbols is in $\mathcal{D}$. The probability of the latter event is

$$\left(1 - Q_Z(\mathcal{D})\right)^n, \tag{A.5}$$

which tends to zero as n tends to infinity because $Q_Z(\mathcal{D}) > 0$.

On the other hand, when transmitting x, an undetected error occurs only if there exist $i \in [n]$ and $x' \in \mathcal{I}' \setminus \{x\}$, such that

$$Z_i \in x' - x + \mathcal{D}. \tag{A.6}$$

Due to (A.1), this implies $Z_i \notin \mathcal{D}$. Thus, by the Union Bound, the probability of error is upper bounded by

$$\sum_{i \in [n]} Q_Z\left(\left(\bigcup_{x' \in \mathcal{I}'} x' - x + \mathcal{D}\right) \setminus \mathcal{D}\right), \tag{A.7}$$

which equals zero because each summand is the probability of a countable union of events of probability zero: $x' - x + \mathcal{D}$ is countable for each x', so is their union over x', and $Q_Z(\{z\}) = 0$ for all $z \notin \mathcal{D}$. $\square$

A.2   Proof of Claim 3.8

Fix $\epsilon > 0$. Let $\mathsf{R} \triangleq \mathsf{C} - 3\epsilon$ and $\mathcal{M} = \{1, \ldots, 2^{n\mathsf{R}}\}$. Fix P_X with $\mathbb{E}_{P_X}[\mathsf{g}(X)] < \Gamma$ and $P_X([\mathsf{a}, \mathsf{b}]) = 1$, such that the induced joint distribution P_{XY} satisfies $I(X;Y) > \mathsf{C} - \epsilon$. Thus,

$$\mathsf{R} + 2\epsilon < I(X;Y). \tag{A.8}$$

Denote the Y-marginal distribution of P_{XY} by P_Y. Then, because Q_Z has a density, $P_Y = P_X * Q_Z$ also has a density, which we denote by f_Y.

Generate $|\mathcal{M}|$ codewords $X^n(1), X^n(2), \ldots, X^n(|\mathcal{M}|)$ independently, each according to the product distribution P_X^n. If message $m \in \mathcal{M}$ is transmitted, the decoder ϕ_{Th} errs if, and only if, the event

$$E_1 = \Big[\mathsf{g}(X^n(m)) > \Gamma \Big] \tag{A.9}$$

or the event

$$E_2 = \Big[\max_{m' \in \mathcal{M} \setminus \{m\}} f_Z^n(Y^n - X^n(m')) \geq 2^{-n\epsilon} f_Z^n(Y^n - X^n(m)) \Big] \tag{A.10}$$

occurs.

The probability of E_1 vanishes by WLLN because $\mathbb{E}_{P_X}[\mathsf{g}(X)] < \Gamma$ by assumption. It essentially follows from [33, Lemma 2] that if (A.8) holds, the probability of E_2 also vanishes. To see why, fix $\{\delta_n\} \downarrow 0$ tending to zero subexponentially, then for each n,

$$\mathbb{P}\Big[\max_{m' \in \mathcal{M} \setminus \{m\}} f_Z(Y^n - X^n(m')) \geq 2^{-n\epsilon} f_Z(Y^n - X^n(m)) \Big]$$

$$= \mathbb{P}\Big[\max_{m' \in \mathcal{M} \setminus \{m\}} \frac{f_Z(Y^n - X^n(m'))}{f_Y(Y^n)} \geq 2^{-n\epsilon} \frac{f_Z(Y^n - X^n(m))}{f_Y(Y^n)} \Big] \tag{A.11}$$

$$\leq \mathbb{P}\Big[\max_{m' \in \mathcal{M} \setminus \{m\}} \frac{f_Z(Y^n - X^n(m'))}{f_Y(Y^n)} > \frac{|\mathcal{M}|}{\delta_n} \Big]$$

$$+ \mathbb{P}\Big[\frac{f_Z(Y^n - X^n(m))}{f_Y(Y^n)} < \frac{2^{n\epsilon} |\mathcal{M}|}{\delta_n} \Big]. \tag{A.12}$$

The first term in the RHS of (A.12) can be upper bounded as

$$\mathbb{P}\Big[\max_{m' \in \mathcal{M} \setminus \{m\}} \frac{f_Z(Y^n - X^n(m'))}{f_Y(Y^n)} > \frac{|\mathcal{M}|}{\delta_n} \Big]$$

$$\leq |\mathcal{M}| \cdot \mathbb{P}\Big[\frac{f_Z(Y^n - X^n(m'))}{f_Y(Y^n)} > \frac{|\mathcal{M}|}{\delta_n} \Big] \tag{A.13}$$

$$\leq |\mathcal{M}| \cdot \frac{\mathbb{E}_{Y^n \sim P_Y^n}\big[\mathbb{E}_{X^n(m') \sim P_X^n}[f_Z(Y^n - X^n(m'))] \big]}{\mathbb{E}_{Y^n \sim P_Y^n}\big[f_Y(Y^n) \big]} \cdot \frac{\delta_n}{|\mathcal{M}|} \tag{A.14}$$

$$= \delta_n, \tag{A.15}$$

which tends to zero as n tends to infinity. Here, (A.13) follows from the Union Bound; in (A.14) we apply Markov's inequality and use the fact that $(X^n(m'), Y^n)$ are generated according to the product distribution

$P_X^n P_Y^n$; and (A.15) holds because for any $y^n \in \mathbb{R}$,

$$f_Y^n(y^n) = \frac{\mathrm{d}P_Y^n}{\mathrm{d}\nu}(y^n) \tag{A.16}$$

$$= \frac{\mathrm{d}(P_Z^n \star P_X^n)}{\mathrm{d}\nu}(y^n) \tag{A.17}$$

$$= \left(\frac{\mathrm{d}P_Z^n}{\mathrm{d}\nu} \star P_X^n\right)(y^n) \tag{A.18}$$

$$= \int f_Z^n(y^n - x^n)\,\mathrm{d}P_X^n(x^n), \tag{A.19}$$

where ν denotes the Lebesgue measure on $\mathbb{R}^n$.

The second term in (A.12) can be rewritten as

$$\mathbb{P}\left[\frac{f_Z(Y^n - X^n(m))}{f_Y(Y^n)} < \frac{2^{n\epsilon}|\mathcal{M}|}{\delta_n}\right]$$

$$= \mathbb{P}\left[\frac{1}{n}\sum_{k=1}^{n}\log\frac{f_Z(Y_k - X_k(m))}{f_Y(Y_k)} < \epsilon + \frac{1}{n}\log\frac{|\mathcal{M}|}{\delta_n}\right] \tag{A.20}$$

$$\leq \mathbb{P}\left[\frac{1}{n}\sum_{k=1}^{n}\log\frac{f_Z(Y_k - X_k(m))}{f_Y(Y_k)} < \mathsf{R} + 2\epsilon\right], \tag{A.21}$$

where (A.21) holds for sufficiently large n, because δ_n decays to zero subexponentially, and thus,

$$\frac{1}{n}\log\frac{|\mathcal{M}|}{\delta_n} = \mathsf{R} - \frac{\log\delta_n}{n} < \mathsf{R} + \epsilon \tag{A.22}$$

for sufficiently large n. Because $(X_1(m), Y_1), \ldots, (X_n(m), Y_n) \sim$ IID P_{XY}, and by assumption,

$$I(X;Y) = \mathbb{E}_{(X,Y)\sim P_{XY}}\left[\log\frac{f_Z(Y - X)}{f_Y(Y)}\right] > \mathsf{R} + 2\epsilon, \tag{A.23}$$

we can apply WLLN to conclude that the RHS of (A.21) vanishes as n tends to infinity. Thus, the probability of event E_2 also vanishes.

By the Union Bound over E_1 and E_2, this proves that the average probability of error (over the messages and the codebooks) vanishes. Abandoning the worse half of the messages, we establish the existence of a sequence of codebooks with message sets $\mathcal{M} = \{1, \ldots, 2^{n\mathsf{R}-1}\}$, such that the maximal probability of error under ϕ_{Th} vanishes. For sufficiently large n,

$$\frac{1}{n}\log|\mathcal{M}| \geq \mathsf{R} - \epsilon = \mathsf{C} - 4\epsilon. \tag{A.24}$$

$\square$

Appendix B

Proofs for Chapter 4

B.1 Proof of Lemma 4.1

To evaluate the variational expression

$$\mathsf{R}^{(\rho)}_{\text{cutoff}} = \max_{P_X} \frac{\mathsf{E}_0(\rho, P_X)}{\rho}, \tag{B.1}$$

on the MMANC, we use the following optimality condition.

Lemma B.1 ([37, Theorem 5.6.5]). *On any DMC, a necessary and sufficient condition for a PMF P_X to maximize $\mathsf{E}_0(\rho, P_X)$ is*

$$\sum_{y \in \mathcal{Y}} Q_{Y|X}(y|x)^{\frac{1}{1+\rho}} \, \alpha_y(P_X)^\rho \geq \sum_{y \in \mathcal{Y}} \alpha_y(P_X)^{1+\rho} \tag{B.2}$$

for all $x \in \mathcal{X}$, with equality if $P_X(x) > 0$. Here,

$$\alpha_y(P_X) = \sum_{x \in \mathcal{X}} P_X(x) \, Q_{Y|X}(y|x)^{\frac{1}{1+\rho}}. \tag{B.3}$$

Take $Q_X = \text{Unif}(\mathcal{A})$. For any $y \in \mathcal{A}$, we evaluate

$$\alpha_y(Q_X) = \sum_{x \in \mathcal{A}} Q_X(x) \, Q_{Y|X}(y|x)^{\frac{1}{1+\rho}} \tag{B.4}$$

$$= \sum_{x \in \mathcal{A}} \frac{1}{|\mathcal{A}|} \, Q_{Y|X}(y|x)^{\frac{1}{1+\rho}} \tag{B.5}$$

$$= \frac{1}{|\mathcal{A}|} \sum_{z \in \mathcal{A}} Q_Z(z)^{\frac{1}{1+\rho}}, \tag{B.6}$$

which is irrelevant to y. Hence,

$$\sum_{y \in \mathcal{Y}} Q_{Y|X}(y|x)^{\frac{1}{1+\rho}} \, \alpha_y(Q_X)^{\rho} = \sum_{z \in \mathcal{A}} Q_Z(z)^{\frac{1}{1+\rho}} \, \alpha_y(Q_X)^{\rho} \tag{B.7}$$

$$= \frac{1}{|\mathcal{A}|^{\rho}} \left(\sum_{z \in \mathcal{A}} Q_Z(z)^{\frac{1}{1+\rho}} \right)^{1+\rho} \tag{B.8}$$

$$= |\mathcal{A}| \left(\frac{1}{|\mathcal{A}|} \sum_{z \in \mathcal{A}} Q_Z(z)^{\frac{1}{1+\rho}} \right)^{1+\rho} \tag{B.9}$$

$$= \sum_{y \in \mathcal{Y}} \alpha_y(Q_X)^{1+\rho}, \tag{B.10}$$

which implies that the uniform distribution maximizes $\mathsf{E}_0(\rho, P_X)$. Thus,

$$\mathsf{R}^{(\rho)}_{\text{cutoff}} = \frac{\mathsf{E}_0(\rho, Q_X)}{\rho} \tag{B.11}$$

$$= -\frac{1}{\rho} \log \sum_{y \in \mathcal{Y}} \left(\frac{1}{|\mathcal{A}|} \cdot \sum_{z \in \mathcal{A}} Q_Z(z)^{\frac{1}{1+\rho}} \right)^{1+\rho} \tag{B.12}$$

$$= \log |\mathcal{A}| - H_{\frac{1}{1+\rho}}(Q_Z). \tag{B.13}$$

$\square$

B.2 Proof of Theorem 4.5

Assume that $0 \leq \mathsf{R}_{\text{h}} < H_{\tilde{\rho}}(Q_Z)$. We only need to prove that

$$\mathsf{R}^{(\rho)}_{\text{cutoff,both,F}} \leq \log |\mathcal{A}| - H_{\tilde{\rho}}(Q_Z) + \mathsf{R}_{\text{h}}. \tag{B.14}$$

Fix any sequence of rate-R_{h} helpers and fix some $t \in \mathcal{T}$. Conditional on $T = t$, the channel is the modulo-additive noise channel on $\mathcal{A}^n$ of law

$$Q_{\mathbf{Y}|\mathbf{X}, T=t}(\mathbf{y}|\mathbf{x}) = Q_{\mathbf{Z}|T=t}(\mathbf{y} \ominus \mathbf{x}), \qquad \mathbf{x}, \mathbf{y} \in \mathcal{A}^n. \tag{B.15}$$

Fix any sequence of rate-R length-n feedback encoders $\mathbf{f}_m = (\mathsf{f}_{m,1}, \ldots, \mathsf{f}_{m,n})$, with

$$\mathsf{f}_{m,k} \colon \mathcal{A}^{k-1} \to \mathcal{A} \tag{B.16a}$$

$$y^{k-1} \mapsto x_k = x_k(m, t, y^{k-1}) \tag{B.16b}$$

for each $k \in [n]$, where t, as above, is fixed.

We adopt the trick as in [23, Section II]: Instead of guessing the message m, think of the decoder guessing the function $\mathbf{f}_m$ based on y^n (and the fixed constant t). Let $\mathcal{F}$ denote the collection of all the n-tuples of mappings $\mathbf{f} = (f_1, \ldots, f_n)$ with $f_k \colon \mathcal{A}^{k-1} \to \mathcal{A}$. Define the PMF on $\mathcal{F}$

$$\tilde{Q}_{\mathbf{F}}(\mathbf{f}) = \frac{|\{m \colon \mathbf{f}_m = \mathbf{f}\}|}{2^{n\mathsf{R}}}, \tag{B.17}$$

and the likelihood function is given by

$$\tilde{Q}_{\mathbf{Y}|\mathbf{F}}(\mathbf{y}|\mathbf{f}) = Q_{\mathbf{Y}|\mathbf{X},T=t}(\mathbf{y}|\mathbf{f}(\mathbf{y})) \tag{B.18}$$

with $\mathbf{f}(\mathbf{y}) = (f_1(y^0), f_2(y^1), \ldots, f_n(y^{n-1})) \in \mathcal{A}^n$.

List the functions $\{\mathbf{f}_m\}$ in decreasing order of likelihood function and denote the ranking of the encoding function $\mathbf{f}_m$ in the list by $\mathsf{G}(m|y^n, t)$. Note that, because guessing a message is equivalent to guessing the encoding function, we still have

$$|\mathcal{L}(m, y^n, t)| \geq \mathsf{G}(m|y^n, t). \tag{B.19}$$

Arıkan's inequality on guessing implies

$$\begin{aligned}
\mathbb{E}&\left[\mathsf{G}(M|Y^n, t)^\rho \,\middle|\, T = t\right] \\
&\geq (1 + n\mathsf{R})^{-\rho}\, 2^{n\rho\mathsf{R} - \mathsf{E}_0(\rho, \tilde{Q}_{\mathbf{F}}, \tilde{Q}_{\mathbf{Y}|\mathbf{F}})} \tag{B.20} \\
&\geq (1 + n\mathsf{R})^{-\rho}\, 2^{n\rho\mathsf{R} - \max_{\tilde{P} \in \mathcal{P}(\mathcal{F})} \mathsf{E}_0(\rho, \tilde{P}, \tilde{Q}_{\mathbf{Y}|\mathbf{F}})}, \tag{B.21}
\end{aligned}$$

where

$$\mathsf{E}_0(\rho, \tilde{P}, \tilde{Q}_{\mathbf{Y}|\mathbf{F}}) = -\log \sum_{\mathbf{y} \in \mathcal{A}^n} \left(\sum_{\mathbf{f} \in \mathcal{F}} \tilde{P}(\mathbf{f})\, \tilde{Q}_{\mathbf{Y}|\mathbf{F}}(\mathbf{y}|\mathbf{f})^{\frac{1}{1+\rho}} \right)^{1+\rho}. \tag{B.22}$$

To analyze (B.21), we claim the following.

Claim B.2.

$$\max_{\tilde{P} \in \mathcal{P}(\mathcal{F})} \mathsf{E}_0(\rho, \tilde{P}, \tilde{Q}_{\mathbf{Y}|\mathbf{F}}) = n\rho \log |\mathcal{A}| - \rho H_{\tilde{\rho}}(Z^n|T = t). \tag{B.23}$$

Using (B.21) and Claim B.2, the lower bound (4.78) is recovered, from which the converse follows in the same way as in Section 4.2.4. $\quad\square$

Proof of claim B.2. For simplicity of notation, we drop the condition on $T = t$ and use notation $Q_{\mathbf{Z}}$ for $Q_{\mathbf{Z}|T=t}$. We emphasize that the channel is now determined by the non-IID noise distribution

$$Q_{\mathbf{Y}|\mathbf{X}}(\mathbf{y}|\mathbf{x}) = Q_{\mathbf{Z}}(\mathbf{y} \ominus \mathbf{x}), \qquad \mathbf{x}, \mathbf{y} \in \mathcal{A}^n, \tag{B.24}$$

and thus, the analysis in [23, Section II] does not carry over directly: it holds only on the memoryless channels. Luckily, although the channel has memory, its additive structure comes to our rescue.

Let $\tilde{P}$ be the PMF on $\mathcal{F}$ that is concentrated and uniform over the constant functions:

$$P^\star(\mathbf{f}) \triangleq \begin{cases} |\mathcal{A}|^{-n} & \text{if } \mathbf{f} \equiv \mathbf{x} \text{ for some } \mathbf{x} \in \mathcal{A}^n \\ 0 & \text{otherwise,} \end{cases} \tag{B.25}$$

where $\mathbf{f} \equiv \mathbf{x}$ means that, irrespective of $\mathbf{y} \in \mathcal{A}^n$, $\mathbf{f}(\mathbf{y}) = \mathbf{x}$. We will establish that $P^\star$ achieves the maximum in the LHS of (B.23) and induces the desired value of E_0 function by applying Lemma B.1 to the channel $\tilde{Q}_{\mathbf{Y}|\mathbf{F}}$.

Indeed, for any $\mathbf{y} \in \mathcal{A}^n$,

$$\alpha_{\mathbf{y}}(P^\star) = \sum_{\mathbf{f} \in \mathcal{F}} P^\star(\mathbf{f})\, \tilde{Q}_{\mathbf{Y}|\mathbf{F}}(\mathbf{y}|\mathbf{f})^{\frac{1}{1+\rho}} \tag{B.26}$$

$$= \sum_{\mathbf{x} \in \mathcal{A}^n} \frac{1}{|\mathcal{A}|^n}\, Q_{\mathbf{Y}|\mathbf{X}}(\mathbf{y}|\mathbf{x})^{\frac{1}{1+\rho}} \tag{B.27}$$

$$= \frac{1}{|\mathcal{A}|^n} \sum_{\mathbf{z} \in \mathcal{A}^n} Q_{\mathbf{Z}}(\mathbf{z})^{\frac{1}{1+\rho}} \tag{B.28}$$

$$\triangleq \alpha \tag{B.29}$$

where in the last step we drop the dependency on $\mathbf{y}$.

It can be verified that

$$\sum_{\mathbf{y} \in \mathcal{A}^n} \tilde{Q}_{\mathbf{Y}|\mathbf{F}}(\mathbf{y}|\mathbf{f})^{\frac{1}{1+\rho}}\, \alpha_{\mathbf{y}}(P^\star)^{\rho}$$

$$= \alpha^{\rho} \sum_{\mathbf{y} \in \mathcal{A}^n} \tilde{Q}_{\mathbf{Y}|\mathbf{F}}(\mathbf{y}|\mathbf{f})^{\frac{1}{1+\rho}} \tag{B.30}$$

$$= \alpha^{\rho} \sum_{\mathbf{y} \in \mathcal{A}^n} Q_{\mathbf{Z}}(\mathbf{y} \ominus \mathbf{f}(\mathbf{y}))^{\frac{1}{1+\rho}} \tag{B.31}$$

$$= \alpha^{\rho} \sum_{y_n \in \mathcal{A}} Q_{Z_n|Z^{n-1}}(y_n \ominus \mathsf{f}_n(y^{n-1}))^{\frac{1}{1+\rho}}$$

$$\times \sum_{y_{n-1} \in \mathcal{A}} Q_{Z_{n-1}|Z^{n-2}}(y_{n-1} \ominus \mathsf{f}_{n-1}(y^{n-2}))^{\frac{1}{1+\rho}}$$

$$\times \cdots \times \sum_{y_1 \in \mathcal{A}} Q_{Z_1}(y_1 \ominus \mathsf{f}_1(y^0))^{\frac{1}{1+\rho}} \tag{B.32}$$

$$= \alpha^\rho \sum_{z_n \in \mathcal{A}} Q_{Z_n | Z^{n-1}}(z_n)^{\frac{1}{1+\rho}}$$

$$\times \sum_{z_{n-1} \in \mathcal{A}} Q_{Z_{n-1} | Z^{n-2}}(z_{n-1})^{\frac{1}{1+\rho}}$$

$$\times \cdots \times \sum_{z_1 \in \mathcal{A}} Q_{Z_1}(z_n)^{\frac{1}{1+\rho}} \tag{B.33}$$

$$= \alpha^\rho \sum_{z^n \in \mathcal{A}^n} Q_{\mathbf{z}}(\mathbf{z})^{\frac{1}{1+\rho}} \tag{B.34}$$

$$= |\mathcal{A}|^n \, \alpha^{1+\rho} \tag{B.35}$$

$$= \sum_{\mathbf{y} \in \mathcal{A}^n} \alpha_{\mathbf{y}}(P^\star)^{1+\rho}. \tag{B.36}$$

This implies that $P^\star$ maximizes the E_0 function. It induces

$$\mathsf{E}_0(\rho, P^\star, Q_{\mathbf{Y}|\mathbf{F}}) = -\log \sum_{\mathbf{y} \in \mathcal{A}^n} \left(\sum_{\mathbf{f} \in \mathcal{F}} P^\star(\mathbf{f}) \, Q_{\mathbf{Y}|\mathbf{F}}(\mathbf{y}|\mathbf{f})^{\frac{1}{1+\rho}} \right)^{1+\rho} \tag{B.37}$$

$$= -\log \sum_{\mathbf{y} \in \mathcal{A}^n} \left(\sum_{\mathbf{z} \in \mathcal{A}^n} \frac{1}{|\mathcal{A}|^n} Q_{\mathbf{z}}(\mathbf{z})^{\frac{1}{1+\rho}} \right)^{1+\rho} \tag{B.38}$$

$$= n\rho \log |\mathcal{A}| - \rho H_{\frac{1}{1+\rho}}(\mathbf{Z}). \tag{B.39}$$

$$\square$$

B.3   Existence of Joint Type $P_{Z,U}$ and Codebook $\mathcal{C}_U$

We state and prove the following lemma, which establishes the existence of the joint type $P_{Z,U} \in \mathcal{P}_n(\mathcal{A} \times \{0,1\})$ subject to the properties required in Section 4.2.2.

Lemma B.3. *Fix type* $P_Z \in \mathcal{P}_n(\mathcal{A})$, $q_0 \in (0,1)$, *and* $\epsilon, \delta > 0$, *there exists some* $n_0(|\mathcal{A}|, q_0, \epsilon, \delta)$ *large enough such that for any* $n \geq n_0$ *and* $q \in [q_0, 1)$ *with* $nq \in \mathbb{N}$, *there exists a joint type* $P_{Z,U} \in \mathcal{P}_n(\mathcal{A} \times \{0,1\})$, *whose P-marginal is* P_Z, *whose U marginal is* $\mathrm{Ber}(q)$, *and which satisfies*

$$I(P_Z, P_{U|Z}) < \epsilon, \tag{B.40}$$

and

$$\left| H(P_{Z|U=1}) - H(P_Z) \right| < \delta. \tag{B.41}$$

Using the above result with $q_0 = \mathsf{R_h}/H(\hat{P}_{Z^n})$ and $q \triangleq \lceil n\mathsf{R_h}/H(\hat{P}_{Z^n}) \rceil / n$, for sufficiently large n, we can find a joint type $P_{Z,U}$ with P-marginal $\hat{P}_{Z^n}$ and U marginal $\mathrm{Ber}(q)$ which satisfies (B.40) and (B.41). The Type Covering Lemma [26, Lemma 3.34] then guarantees that we can find a collection $\mathcal{C}_U \subset \{0,1\}^n$ of cardinality

$$|\mathcal{C}_U| < 2^{n(I(P_Z, P_{U|Z})+\epsilon)} < 2^{2n\epsilon}, \tag{B.42}$$

such that for any $z^n \in \mathcal{T}^{(n)}_{\hat{P}_{Z^n}}$, there exists $u^n \in \mathcal{C}_U$ whose the joint type with z^n is exactly $P_{Z,U}$. Simply substituting ϵ by $\epsilon/2$ yields the existence of the codebook we want.

To establish the lemma, note that, indeed, ignoring the requirement that $P_{Z,U}$ is a type of denominator n, we can let Z and U independent. Since types become dense when n tends to infinity, the effects of rounding also become negligible. The technical details can be found below.

Proof of Lemma B.3. Recall that $\mathcal{A} = \{0, 1, \ldots, |\mathcal{A}| - 1\}$. Without loss of generality, assume $P_Z(i) > 0$, for all $i \in \mathcal{A}$ (otherwise shrink the alphabet).

Given $P_Z \in \mathcal{P}_n(\mathcal{A})$, to construct $P_{Z,U}$, we shall fix some $z^n \in \mathcal{T}^{(n)}_{P_Z}$, construct correspondingly $u^n \in \{0,1\}^n$, and set $P_{Z,U}$ as the empirical type $\hat{P}_{(z^n, u^n)}$. For $i \in \mathcal{A} \setminus \{0\}$, among the $n_i = nP_Z(i)$ coordinates in z^n that corresponds to symbol i, we pick $n_{i,1} = \lceil n_i q \rceil$ of them to be one in u^n, and the remaining $n_{i,0} = n_i - n_{i,1}$ zero. For $i = 0$, among the remaining $n_0 = nP_Z(0)$ coordinates, we pick

$$n_{0,1} = nq - \sum_{i \in \mathcal{A} \setminus \{0\}} n_{i,1} \tag{B.43}$$

of them to be one, and the other $n_{0,0} = n_0 - n_{0,1}$ coordinates zero. Note that, this construction is valid since

$$n_{0,1} = nq - \sum_{i \in \mathcal{A} \setminus \{0\}} \lceil n_i q \rceil \tag{B.44}$$

$$> nq - \sum_{i \in \mathcal{A} \setminus \{0\}} n_i q - |\mathcal{A}| + 1 \tag{B.45}$$

$$= nqP_Z(0) - |\mathcal{A}| + 1 \tag{B.46}$$

$$\geq 0, \tag{B.47}$$

where (B.47) holds for n sufficiently because by assumption $qP_Z(0) > 0$.

The construction guarantees that the type of u^n is $\mathrm{Ber}(q)$, and we shall establish that, for sufficiently large n, it also leads to a joint type $P_{Z,U}$ satisfying (B.40) and (B.41). To that end, we cite the following lemma on the continuity of entropy.

Lemma B.4 ([63]). *If the total variation distance of two distributions P and Q on $\mathcal{X}$ is*

$$\Theta \triangleq d(P,Q) = \sum_{x \in \mathcal{X}} |P(x) - Q(x)|, \tag{B.48}$$

and $\Theta < 1/2$, then

$$|H(P) - H(Q)| \leq -\Theta \log \frac{\Theta}{|\mathcal{X}|}. \tag{B.49}$$

Note that, for $i \in \mathcal{A} \setminus \{0\}$

$$P_{Z|U=1}(i) \in \left[P_Z(i), P_Z(i) + \frac{1}{nq} \right), \tag{B.50}$$

and

$$P_{Z|U=1}(0) \in \left(P_Z(0) - \frac{|\mathcal{A}| - 1}{nq}, P_Z(0) \right], \tag{B.51}$$

Thus, the total variation distance satisfies

$$\Theta_1 \triangleq d(P_{Z|U=1}, P_Z) = \sum_{i \in \mathcal{A}} |P_{Z|U=1}(i) - P_Z(i)| < \frac{2|\mathcal{A}|}{nq}, \tag{B.52}$$

and similarly,

$$\Theta_0 \triangleq d(P_{Z|U=0}, P_Z) < \frac{2|\mathcal{A}|}{n(1 - q)}. \tag{B.53}$$

Therefore, using Lemma B.4, we have

$$\left| H(P_{Z|U=1}) - H(P_Z) \right| \leq -\Theta_1 \log \frac{\Theta_1}{|\mathcal{A}|} \tag{B.54}$$

$$< \frac{2|\mathcal{A}|}{nq} \log \frac{nq}{2} \tag{B.55}$$

$$< \frac{2|\mathcal{A}|}{nq_0} \log \frac{n}{2} \tag{B.56}$$

and

$$I(P_Z, P_{U|Z}) = H(P_Z) - qH(P_{Z|U=1}) - (1-q)H(P_{Z|U=0}) \tag{B.57}$$

$$\leq -q\Theta_1 \log \frac{\Theta_1}{|\mathcal{A}|} - (1-q)\Theta_0 \log \frac{\Theta_0}{|\mathcal{A}|} \tag{B.58}$$

$$< \frac{2|\mathcal{A}|}{n} \log \frac{n^2 q(1-q)}{4} \tag{B.59}$$

$$\leq \frac{2|\mathcal{A}|}{n} \log \frac{n^2}{16} \tag{B.60}$$

where (B.55) and (B.59) hold for sufficiently large n because $t \mapsto -t \log t$ is monotonically increasing on $(0, 1/e)$. Both (B.56) and (B.60) tends to zero as $n \to \infty$, thus (B.40) and (B.41) holds for sufficiently large n, depending on $|\mathcal{A}|$, q_0, ϵ, and δ. $\qquad\square$

B.4　Proof of Lemma 4.21

We apply Proposition 4.20 with the density f_{R} corresponding to a centered Gaussian of variance σ^2, where

$$\sigma^2 = \frac{\mathsf{A}}{(1+\rho)\beta} + 1 \tag{B.61}$$

and

$$\beta = \frac{1}{2}\left(1 - \frac{\mathsf{A}}{1+\rho} + \sqrt{\left(1 - \frac{\mathsf{A}}{1+\rho}\right)^2 + \frac{4\mathsf{A}}{(1+\rho)^2}}\right). \tag{B.62}$$

Evaluating the RHS of (4.122) for this density, we obtain

$$\mathsf{E}_0^{(n)}(\rho, Q^{(n)})$$

$$\leq \sup_{P:\, \mathbb{E}[X^2] \leq \mathsf{A}} -(1+\rho) \int_{x \in \mathbb{R}} \ln \int_{y \in \mathbb{R}} w(y|x)^{\frac{1}{1+\rho}} f_{\mathrm{R}}(y)^{\frac{\rho}{1+\rho}} \, \mathrm{d}y \, \mathrm{d}P(x) \tag{B.63}$$

$$= \sup_{P:\, \mathbb{E}[X^2] \leq \mathsf{A}} -(1+\rho) \int_{x \in \mathbb{R}} \ln \int_{y \in \mathbb{R}} \frac{1}{(\sqrt{2\pi})^{\frac{1}{1+\rho}}} e^{-\frac{(y-x)^2}{2(1+\rho)}}$$

$$\cdot \frac{1}{(\sqrt{2\pi\sigma^2})^{\frac{\rho}{1+\rho}}} e^{-\frac{y^2\rho}{2\sigma^2(1+\rho)}} \, \mathrm{d}y \, \mathrm{d}P(x) \tag{B.64}$$

$$= \sup_{P:\,\mathbb{E}[X^2]\leq A} -(1+\rho)\int_{x\in\mathbb{R}}\ln\left(\sqrt{\frac{2\pi(1+\rho)^2}{\rho}}\sqrt{\sigma^{2}}^{\frac{1}{1+\rho}}\right)$$
$$\cdot\frac{1}{\sqrt{2\pi\sigma_1^2}}e^{-\frac{x^2}{2\sigma_1^2}}\,\mathrm{d}P(x)\quad\text{(B.65)}$$

$$= \sup_{P:\,\mathbb{E}[X^2]\leq A} -(1+\rho)\ln\left(\sqrt{\frac{(1+\rho)^2}{\rho}}\sqrt{\sigma^{2}}^{\frac{1}{1+\rho}}\frac{1}{\sqrt{\sigma_1^2}}\right)$$
$$+(1+\rho)\int_{x\in\mathbb{R}}\frac{x^2}{2\sigma_1^2}\,\mathrm{d}P(x)\quad\text{(B.66)}$$

$$= -(1+\rho)\ln\left(\sqrt{\frac{(1+\rho)^2}{\rho}}\sqrt{\sigma^{2}}^{\frac{1}{1+\rho}}\frac{1}{\sqrt{\sigma_1^2}}\right)+(1+\rho)\frac{A}{2\sigma_1^2}\quad\text{(B.67)}$$

$$= \frac{1+\rho}{2}\frac{A}{\sigma_1^2}+\frac{1+\rho}{2}\ln\sigma_1^2-\frac{1}{2}\ln\sigma^2-\frac{1+\rho}{2}\ln\frac{(1+\rho)^2}{\rho}\quad\text{(B.68)}$$

where in (B.65) we defined

$$\sigma_1^2 \triangleq 1+\rho+\frac{\sigma^2(1+\rho)}{\rho}\quad\text{(B.69a)}$$
$$= \frac{A}{\rho\beta}+\frac{(1+\rho)^2}{\rho}.\quad\text{(B.69b)}$$

To conclude the proof, it remains to show that the RHS of (B.68) coincide with $\rho R_0^{(\rho)}$. To this end, observe that some basic algebra reveals that

$$\beta\left(\beta-1+\frac{A}{1+\rho}\right)=\frac{A}{(1+\rho)^2}\quad\text{(B.70)}$$

and

$$\left(\beta+\frac{A}{1+\rho}\right)(1-\beta)=\frac{A\rho}{(1+\rho)^2}.\quad\text{(B.71)}$$

Therefore, the first term in (B.68) can be rewritten as

$$\frac{1+\rho}{2}\frac{A}{\sigma_1^2}=\frac{1+\rho}{2}\frac{A}{\frac{A}{\rho\beta}+\frac{(1+\rho)^2}{\rho}}=\frac{1+\rho}{2}\frac{A\rho}{(1+\rho)^2}\frac{\beta}{\frac{A}{(1+\rho)^2}+\beta}\quad\text{(B.72)}$$
$$=\frac{A\rho}{2(1+\rho)}\frac{1}{\beta+\frac{A}{1+\rho}}=\frac{(1+\rho)(1-\beta)}{2},\quad\text{(B.73)}$$

and the remaining terms rewritten as

$$\frac{1+\rho}{2}\ln\sigma_1^2 - \frac{1}{2}\ln\sigma^2 - \frac{1+\rho}{2}\ln\frac{(1+\rho)^2}{\rho}$$

$$= \frac{1+\rho}{2}\ln\left(\frac{\mathsf{A}}{\rho\beta} + \frac{(1+\rho)^2}{\rho}\right) - \frac{1}{2}\ln\left(\frac{\mathsf{A}}{(1+\rho)\beta} + 1\right)$$

$$- \frac{1+\rho}{2}\ln\frac{(1+\rho)^2}{\rho} \tag{B.74}$$

$$= \frac{1+\rho}{2}\ln\left(\frac{\mathsf{A}}{(1+\rho)^2\beta} + 1\right) - \frac{1}{2}\ln\left(\frac{\mathsf{A}}{(1+\rho)\beta} + 1\right) \tag{B.75}$$

$$= \frac{\rho}{2}\ln\left(\beta + \frac{\mathsf{A}}{1+\rho}\right) + \frac{1}{2}\ln\beta. \tag{B.76}$$

The sum equals $\rho\mathsf{R}_0^{(\rho)}$ as in (4.92). $\qquad\qquad\qquad\qquad\qquad\qquad\square$

B.5   Proof of Lemma 4.22

The implication "$\Longleftarrow$" follows by noting for any $z \in \mathbb{N}$ and $\rho > 0$

$$(1+z)^\rho \leq 1 + 2^\rho z^\rho, \tag{B.77}$$

so

$$\mathbb{E}\left[(1+Z_n)^\rho\right] \leq 1 + 2^\rho\,\mathbb{E}\left[Z_n^\rho\right]. \tag{B.78}$$

As for the implication "$\Longrightarrow$", note that for any $z \in \mathbb{N}$ and $\rho > 0$

$$(1+z)^\rho \geq z^\rho + \mathbb{1}\{z = 0\}, \tag{B.79}$$

so

$$\mathbb{E}\left[(1+Z_n)^\rho\right] \geq \mathbb{E}\left[Z_n^\rho\right] + \mathbb{P}[Z_n = 0]. \tag{B.80}$$

The implication is now established by noting that the LHS of (4.139) implies that $\lim_{n\to\infty}\mathbb{P}[Z_n = 0] = 1$ because, by Markov's inequality (and the strict positivity of ρ),

$$\mathbb{P}[Z_n \neq 0] = \mathbb{P}\left[(1+Z_n)^\rho - 1 \geq 2^\rho - 1\right] \tag{B.81}$$

$$\leq \frac{\mathbb{E}\left[(1+Z_n)^\rho\right] - 1}{2^\rho - 1}. \tag{B.82}$$

$$\square$$

B.6 Proof of Lemma 4.23

We shall establish that the expectation

$$\mathbb{E}\left[\mathbb{E}\left[\sum_{m\neq 1} B_m(\mathbf{x}_1,\mathbf{y})\,\middle|\,\mathbf{X}_1=\mathbf{x}_1,\mathbf{Y}=\mathbf{y}\right]^{\rho}\right]$$

$$=\int_{\mathbf{y}\in\mathbb{R}^n}\int_{\mathbf{x}_1\in\mathbb{R}^n}\mathbb{E}\left[\sum_{m\neq 1} B_m(\mathbf{x}_1,\mathbf{y})\,\middle|\,\mathbf{X}_1=\mathbf{x}_1,\mathbf{Y}=\mathbf{y}\right]^{\rho}$$

$$\cdot w(\mathbf{y}|\mathbf{x}_1)\,\mathrm{d}Q^{(n)}(\mathbf{x}_1)\,\mathrm{d}\nu(\mathbf{y}) \tag{B.83}$$

tends to zero as n tends to infinity whenever $\mathsf{R}<\mathsf{R}_0^{(\rho)}$.

First notice that conditional on the transmitted codeword $\mathbf{x}_1$ and the channel output $\mathbf{y}$, the random variables $\{B_m\}_{m\neq 1}$ are IID Bernoulli, with B_m determined by $\mathbf{X}_m$ and being of probability of success

$$p(\mathbf{x}_1,\mathbf{y})=\mathbb{P}\left[w(\mathbf{y}|\mathbf{X}_m)\geq w(\mathbf{y}|\mathbf{x}_1)\right] \tag{B.84}$$

$$=\mathbb{P}\left[w(\mathbf{y}|\mathbf{X}_m)^{\frac{1}{1+\rho}}\geq w(\mathbf{y}|\mathbf{x}_1)^{\frac{1}{1+\rho}}\right] \tag{B.85}$$

$$\leq w(\mathbf{y}|\mathbf{x}_1)^{-\frac{1}{1+\rho}}\,\mathbb{E}\left[w(\mathbf{y}|\mathbf{X}_m)^{\frac{1}{1+\rho}}\right], \tag{B.86}$$

where the last inequality follows from Markov's inequality.

Thus,

$$\mathbb{E}\left[\mathbb{E}\left[\sum_{m\neq 1} B_m(\mathbf{x}_1,\mathbf{y})\,\middle|\,\mathbf{X}_1=\mathbf{x}_1,\mathbf{Y}=\mathbf{y}\right]^{\rho}\right]$$

$$=\int_{\mathbf{y}\in\mathbb{R}^n}\int_{\mathbf{x}_1\in\mathbb{R}^n}\mathbb{E}\left[\sum_{m\neq 1} B_m(\mathbf{x}_1,\mathbf{y})\,\middle|\,\mathbf{X}_1=\mathbf{x}_1,\mathbf{Y}=\mathbf{y}\right]^{\rho}$$

$$\cdot w(\mathbf{y}|\mathbf{x}_1)\,\mathrm{d}Q^{(n)}(\mathbf{x}_1)\,\mathrm{d}\nu(\mathbf{y}) \tag{B.87}$$

$$\leq e^{n\rho\mathsf{R}}\int_{\mathbf{y}\in\mathbb{R}^n}\int_{\mathbf{x}_1\in\mathbb{R}^n} p(\mathbf{x}_1,\mathbf{y})^{\rho}\,w(\mathbf{y}|\mathbf{x}_1)\,\mathrm{d}Q^{(n)}(\mathbf{x}_1)\,\mathrm{d}\nu(\mathbf{y}) \tag{B.88}$$

$$\leq e^{n\rho\mathsf{R}}\int_{\mathbf{y}\in\mathbb{R}^n}\int_{\mathbf{x}_1\in\mathbb{R}^n}\mathbb{E}\left[w(\mathbf{y}|\mathbf{X}_m)^{\frac{1}{1+\rho}}\right]^{\rho} w(\mathbf{y}|\mathbf{x}_1)^{\frac{1}{1+\rho}}\,\mathrm{d}Q^{(n)}(\mathbf{x}_1)\,\mathrm{d}\nu(\mathbf{y})$$

$$\tag{B.89}$$

$$=e^{n\rho\mathsf{R}}\int_{\mathbf{y}\in\mathbb{R}^n}\mathbb{E}\left[w(\mathbf{y}|\mathbf{X}_m)^{\frac{1}{1+\rho}}\right]^{\rho}\left(\int_{\mathbf{x}_1\in\mathbb{R}^n} w(\mathbf{y}|\mathbf{x}_1)^{\frac{1}{1+\rho}}\,\mathrm{d}Q^{(n)}(\mathbf{x}_1)\right)\mathrm{d}\nu(\mathbf{y})$$

$$\tag{B.90}$$

$$= e^{n\rho\mathsf{R}} \int_{\mathbf{y}\in\mathbb{R}^n} \left(\int_{\mathbf{x}\in\mathbb{R}^n} w(\mathbf{y}|\mathbf{x})^{\frac{1}{1+\rho}} \, dQ^{(n)}(\mathbf{x}) \right)^{1+\rho} d\nu(\mathbf{y}) \tag{B.91}$$

$$\leq \left(\frac{e^{r\delta}}{\mu}\right)^{1+\rho} e^{n\rho\mathsf{R}} \left(\int_{y\in\mathbb{R}} \left(\int_{x\in\mathbb{R}} e^{r(x^2-\mathsf{A})} w(y|x)^{\frac{1}{1+\rho}} \, dQ_{\mathrm{G}}(x) \right)^{1+\rho} d\nu(y) \right)^n \tag{B.92}$$

$$= \left(\frac{e^{r\delta}}{\mu}\right)^{1+\rho} e^{n\rho\mathsf{R}} \, e^{-n\mathsf{E}_{0,\mathrm{m}}(\rho,Q_{\mathrm{G}},r)}, \tag{B.93}$$

where (B.88) follows from the Linearity of Expectation; and (B.92) follows
from the upper bound (4.114) on the Radon–Nykodim derivative and
holds for every $r \geq 0$. Choosing r as $r^\star$ that achieves $\mathsf{E}^*_{0,\mathrm{m}}(\rho, Q_{\mathrm{G}})$
(c.f. (4.106)), we obtain

$$\mathbb{E}\left[\mathbb{E}\left[\sum_{m\neq 1} B_m(\mathbf{x}_1,\mathbf{y}) \,\middle|\, \mathbf{X}_1 = \mathbf{x}_1, \mathbf{Y} = \mathbf{y} \right]^\rho \right]$$

$$\leq \left(\frac{e^{r^\star\delta}}{\mu}\right)^{1+\rho} e^{n\rho\mathsf{R}} \, e^{-n\mathsf{E}^*_{0,\mathrm{m}}(\rho,Q_{\mathrm{G}})} \tag{B.94}$$

$$= \left(\frac{e^{r^\star\delta}}{\mu}\right)^{1+\rho} e^{n\rho(\mathsf{R}-\mathsf{R}_0^{(\rho)})}, \tag{B.95}$$

where the last equality follows from (4.132).

The Central Limit Theorem guarantees that, as n tends to infinity, μ
approaches $1/2$. Consequently, the RHS of (B.95) tends to zero whenever
$\mathsf{R} < \mathsf{R}_0^{(\rho)}$. $\qquad\square$

B.7 Proof of Lemma 4.24

To prove the lemma, we shall establish that, whenever $\mathsf{R} < \mathsf{R}_0^{(\rho)} - \Delta$,

$$\lim_{n\to\infty} \mathbb{E}\left[\mathbb{E}\left[\sum_{m\neq 1} B_m(\mathbf{x}_1,\mathbf{y};\Delta) \,\middle|\, \mathbf{X}_1 = \mathbf{x}_1, \mathbf{Y} = \mathbf{y} \right]^\rho \right] = 0, \tag{B.96}$$

where the outer expectation is over $\mathbf{X}_1$ and $\mathbf{Y}$. From this (4.175) will
follow in much the same way that (4.140) followed from (4.142) in Section 4.3.2.3.

To establish (B.96), first note that, conditional on the transmitted
codeword $\mathbf{x}_1$ and the channel output $\mathbf{y}$, the random variables $\{B_m(\mathbf{x}_1,\mathbf{y};\Delta)\}_{m\neq 1}$

are IID Bernoulli, with B_m determined by $\mathbf{X}_m$ and being of probability of success

$$p(\mathbf{x}_1, \mathbf{y}; \Delta) = \mathbb{P}\left[w(\mathbf{y}|\mathbf{X}_m) \geq w(\mathbf{y}|\mathbf{x}_1)\, e^{-\frac{n\Delta}{2}}\right] \tag{B.97}$$

$$= \mathbb{P}\left[w(\mathbf{y}|\mathbf{X}_m)^{\frac{1}{1+\rho}} \geq w(\mathbf{y}|\mathbf{x}_1)^{\frac{1}{1+\rho}}\, e^{-\frac{n\Delta}{2(1+\rho)}}\right] \tag{B.98}$$

$$\leq w(\mathbf{y}|\mathbf{x}_1)^{-\frac{1}{1+\rho}}\, e^{\frac{n\Delta}{2(1+\rho)}}\, \mathbb{E}\left[w(\mathbf{y}|\mathbf{X}_m)^{\frac{1}{1+\rho}}\right], \tag{B.99}$$

where the last inequality follows from Markov's inequality. Consequently,

$$\mathbb{E}\left[\mathbb{E}\left[\sum_{m\neq 1} B_m(\mathbf{x}_1, \mathbf{y}; \Delta)\,\bigg|\, \mathbf{X}_1 = \mathbf{x}_1, \mathbf{Y} = \mathbf{y}\right]^{\rho}\right]$$

$$= \int_{\mathbf{y}\in\mathbb{R}^n} \int_{\mathbf{x}_1\in\mathbb{R}^n} \mathbb{E}\left[\sum_{m\neq 1} B_m(\mathbf{x}_1, \mathbf{y}; \Delta)\,\bigg|\, \mathbf{X}_1 = \mathbf{x}_1, \mathbf{Y} = \mathbf{y}\right]^{\rho}$$

$$\cdot w(\mathbf{y}|\mathbf{x}_1)\, \mathrm{d}Q^{(n)}(\mathbf{x}_1)\, \mathrm{d}\nu(\mathbf{y}) \tag{B.100}$$

$$\leq e^{n\rho R} \int_{\mathbf{y}\in\mathbb{R}^n} \int_{\mathbf{x}_1\in\mathbb{R}^n} p(\mathbf{x}_1, \mathbf{y}; \Delta)^{\rho}\, w(\mathbf{y}|\mathbf{x}_1)\, \mathrm{d}Q^{(n)}(\mathbf{x}_1)\, \mathrm{d}\nu(\mathbf{y}) \tag{B.101}$$

$$\leq e^{n\rho R} e^{n\rho\Delta} \int_{\mathbf{y}\in\mathbb{R}^n} \int_{\mathbf{x}_1\in\mathbb{R}^n} \mathbb{E}\left[w(\mathbf{y}|\mathbf{X}_m)^{\frac{1}{1+\rho}}\right]^{\rho} w(\mathbf{y}|\mathbf{x}_1)^{\frac{1}{1+\rho}}\, \mathrm{d}Q^{(n)}(\mathbf{x}_1)\, \mathrm{d}\nu(\mathbf{y})$$

$$\tag{B.102}$$

where (B.102) follows from (B.99) because $\rho, \Delta > 0$, and $n\rho\Delta/(2(1+\rho)) < n\rho\Delta$. Except for the $e^{n\rho\Delta}$ factor, the RHS of (B.102) is identical to the RHS of (B.89), which was shown to decay at least as fast as $e^{n\rho(R-R_0^{(\rho)})}$; see (B.95). It follows that the RHS of (B.102) tends to zero whenever $R + \Delta < R_0^{(\rho)}$. $\qquad\square$

Appendix C

Proofs for Chapter 5

C.1   Proof of Corollary 5.3

We can lower-bound π_0 by lower-bounding the maximum over $y \in \mathcal{A}$ by the average:

$$\pi_0 = \min_{P \in \mathcal{P}(\mathcal{A})} \max_{y \in \mathcal{A}} \sum_{x \in \mathcal{X}_y} P(x) \tag{C.1}$$

$$\geq \frac{1}{|\mathcal{A}|} \min_{P \in \mathcal{P}(\mathcal{A})} \sum_{y \in \mathcal{A}} \sum_{x \in \mathcal{X}_y} P(x) \tag{C.2}$$

$$= \frac{|\mathcal{S}|}{|\mathcal{A}|} \min_{P \in \mathcal{P}(\mathcal{A})} \sum_{x \in \mathcal{A}} P(x) \tag{C.3}$$

$$= \frac{|\mathcal{S}|}{|\mathcal{A}|}, \tag{C.4}$$

where in (C.2) we lower-bounded the maximum over $\mathcal{A}$ by the arithmetic average; and (C.3) holds since in the double sum, each $x \in \mathcal{A}$ is contained in $\mathcal{X}_y$ for exactly $|\mathcal{S}|$ different y's in $\mathcal{A}$. The corollary now follows by noting that this lower bound is tight as can be seen by considering P equiprobable. $\qquad\square$

C.2 Proof of Lemma 5.20

Without loss of generality, assume $R_L > 0$ and $|\mathcal{S}|^n > L$ (otherwise the result is obvious). Define

$$R = \frac{1}{n} \log \lfloor 2^{nR_L} \rfloor \tag{C.5}$$

$$\leq R_L \tag{C.6}$$

and let $\mathcal{M} = \{1, \ldots, 2^{nR}\}$.

Generate a random codebook $\mathcal{C} = \{\mathbf{X}(1), \ldots, \mathbf{X}(|\mathcal{M}|)\}$ by drawing its codewords independently, each equiprobably from $\mathcal{A}^n$. We will show that with positive probability, the properties (i) $\mathcal{C}$ contains no repeating codewords, and (ii) $\max_{\mathbf{z} \in \mathcal{S}^n} |\mathcal{F}_{\mathbf{z}}| \leq L - 1$ hold simultaneously.

Using the Union Bound, we can upper-bound the probability that Property (i) is violated as follows:

$$\mathbb{P}\big[\exists\, m, m' \in \mathcal{M} \text{ s.t. } m \neq m' \text{ and } \mathbf{X}(m) = \mathbf{X}(m')\big]$$

$$\leq \sum_{1 \leq m < m' \leq 2^{nR}} \mathbb{P}\big[\mathbf{X}(m) = \mathbf{X}(m')\big] \tag{C.7}$$

$$= \binom{2^{nR}}{2} \cdot \frac{1}{|\mathcal{A}|^n} \tag{C.8}$$

$$\leq \frac{2^{2nR}}{2|\mathcal{A}|^n}. \tag{C.9}$$

As for Property (ii),

$$\mathbb{P}\big[\max_{\mathbf{z} \in \mathcal{S}^n} |\mathcal{F}_{\mathbf{z}}| > L - 1\big]$$

$$= \mathbb{P}\big[\exists\, \mathbf{z} \in \mathcal{S}^n \text{ s.t. } |\mathcal{F}_{\mathbf{z}}| \geq L\big] \tag{C.10}$$

$$\leq \sum_{\mathbf{z} \in \mathcal{S}^n} \mathbb{P}\big[|\mathcal{F}_{\mathbf{z}}| \geq L\big] \tag{C.11}$$

$$= \sum_{\mathbf{z} \in \mathcal{S}^n} \mathbb{P}\big[\exists\, \text{distinct } \boldsymbol{\xi}_1', \ldots, \boldsymbol{\xi}_L' \in \mathcal{S}^n$$
$$\text{s.t. } \{\boldsymbol{\xi}_1', \ldots, \boldsymbol{\xi}_L'\} \ominus \mathbf{z} \subseteq (\mathcal{C} \ominus \mathcal{C})^\star\big] \tag{C.12}$$

$$= \sum_{\mathbf{z} \in \mathcal{S}^n} \mathbb{P}\big[\exists\, \text{distinct } \boldsymbol{\xi}_1, \ldots, \boldsymbol{\xi}_L \in \mathcal{S}^n \ominus \mathbf{z}$$
$$\text{s.t. } \{\boldsymbol{\xi}_1, \ldots, \boldsymbol{\xi}_L\} \subseteq (\mathcal{C} \ominus \mathcal{C})^\star\big], \tag{C.13}$$

where (C.11) follows from the Union Bound; (C.12) follows from the definition of $\mathcal{F}_{\mathbf{z}}$ in (5.71); and in (C.13) we introduced $\boldsymbol{\xi}_i = \boldsymbol{\xi}_i' \ominus \mathbf{z}$.

To analyze the probabilities appearing on the RHS of (C.13), we first rule out the degenerate cases. We say that a collection of n-tuples $\boldsymbol{\xi}_1, \ldots, \boldsymbol{\xi}_\ell \in \mathcal{A}^n$ is *tri-independent* if

$$\left(\sum_{i=1}^{\ell} \epsilon_i \boldsymbol{\xi}_i = \mathbf{0}, \text{ with } \epsilon_i \in \{0, \pm 1\}, \ \forall\, i \in [\ell] \right)$$
$$\implies \left(\epsilon_i = 0, \ \forall\, i \in [\ell] \right), \tag{C.14}$$

where the summation is calculated with respect to mod-$|\mathcal{A}|$ addition, and $\mathbf{0}$ represents the all-zero sequence,

Lemma C.1. *Among any* L *distinct n-tuples* $\boldsymbol{\xi}_1, \ldots, \boldsymbol{\xi}_\mathsf{L} \in \mathcal{A}^n$, *there exist at least* $\lceil \log_3 \mathsf{L} \rceil$ *that are tri-independent.*

Proof of Lemma C.1. Let $\mathcal{B} \subseteq \{\boldsymbol{\xi}_1, \ldots, \boldsymbol{\xi}_\mathsf{L}\}$ be a maximal tri-independent set (with respect to inclusion), and let ℓ be its cardinality: $|\mathcal{B}| = \ell$. Without loss of generality, assume $\mathcal{B} = \{\boldsymbol{\xi}_1, \ldots, \boldsymbol{\xi}_\ell\}$ (otherwise rearrange the tuples). We will show that every n-tuple in $\{\boldsymbol{\xi}_1, \ldots, \boldsymbol{\xi}_\mathsf{L}\}$ can be expressed in the form $\sum_{i=1}^{\ell} \epsilon_i \boldsymbol{\xi}_i$, with $\epsilon_i \in \{0, \pm 1\}$ for all $i \in [\ell]$. (This is obvious for the n-tuples $\boldsymbol{\xi}_1, \ldots, \boldsymbol{\xi}_\ell$ but requires proof for $\boldsymbol{\zeta}_{\ell+1}, \ldots, \boldsymbol{\xi}_\mathsf{L}$.) This will establish the lemma because the number of distinct expressions of said form is at most 3^ℓ, and the number of sequences is L, so 3^ℓ must be at least L, and ℓ (the number of elements of $\mathcal{B}$) must thus satisfy $\ell \geq \lceil \log_3 \mathsf{L} \rceil$.

To see that any $\boldsymbol{\xi} \in \{\boldsymbol{\xi}_{\ell+1}, \ldots, \boldsymbol{\xi}_\mathsf{L}\}$ can be expressed in said form, note that the maximality of $\mathcal{B}$ implies that for any $\boldsymbol{\xi} \in \{\boldsymbol{\xi}_{\ell+1}, \ldots, \boldsymbol{\xi}_\mathsf{L}\}$, there exist nontrivial (i.e., not all zero) $\{\epsilon_i\}_{i=1}^{\ell}$ and ϵ, such that $\sum_{i=1}^{\ell} \epsilon_i \boldsymbol{\xi}_i + \epsilon \boldsymbol{\xi} = \mathbf{0}$. Here ϵ cannot be zero, since otherwise the relation would translate to $\sum_{i=1}^{\ell} \epsilon_i \boldsymbol{\xi}_i = \mathbf{0}$ for nontrivial $\{\epsilon_i\}_{i=1}^{\ell}$, which would contradict the fact that, by construction, $\mathcal{B}$ is tri-independent. Without loss of generality, assume $\epsilon = -1$ (otherwise change the sign of ϵ and all ϵ_i's), so $\boldsymbol{\xi} = \sum_{i=1}^{\ell} \epsilon_i \boldsymbol{\xi}_i$, thus expressing $\boldsymbol{\xi}$ in said form. $\qquad\square$

With the aid of Lemma C.1, and defining

$$\ell \triangleq \lceil \log_3 \mathsf{L} \rceil \geq 2, \tag{C.15}$$

we can return to the RHS of (C.13) to conclude that

$$\mathbb{P}\left[\exists \text{ distinct } \boldsymbol{\xi}_1, \ldots, \boldsymbol{\xi}_\mathsf{L} \in \mathcal{S}^n \ominus \mathbf{z} \text{ s.t. } \{\boldsymbol{\xi}_1, \ldots, \boldsymbol{\xi}_\mathsf{L}\} \subseteq (\mathcal{C} \ominus \mathcal{C})^\star \right]$$

$$\leq \mathbb{P}\big[\exists \text{ tri-independent } \boldsymbol{\xi}_1,\ldots,\boldsymbol{\xi}_\ell \in \mathcal{S}^n \ominus \mathbf{z} \text{ s.t. } \{\boldsymbol{\xi}_1,\ldots,\boldsymbol{\xi}_\ell\} \subseteq (\mathcal{C}\ominus\mathcal{C})^\star\big]. \tag{C.16}$$

Ignoring, for now, the constraint that $\boldsymbol{\xi}_1,\ldots,\boldsymbol{\xi}_\ell$ be in $\mathcal{S}^n \ominus \mathbf{z}$, we claim:

Claim C.2. *Given ℓ tri-independent n-tuples $\boldsymbol{\xi}_1,\ldots,\boldsymbol{\xi}_\ell \in \mathcal{A}^n$, and a random codebook $\mathcal{C}$ generated as above*

$$\mathbb{P}\big[\{\boldsymbol{\xi}_1,\ldots,\boldsymbol{\xi}_\ell\} \subseteq (\mathcal{C}\ominus\mathcal{C})^\star\big] \leq \left(\frac{2^{2n\mathrm{R}}}{|\mathcal{A}|^n}\right)^\ell. \tag{C.17}$$

Proof. Since $\{\boldsymbol{\xi}_1,\ldots,\boldsymbol{\xi}_\ell\}$ are tri-independent and hence, *a fortiori*, nonzero, the event $\big[\boldsymbol{\xi}_i \in (\mathcal{C}\ominus\mathcal{C})^\star\big]$ is equivalent to the event

$$\big[\exists\,(m_i, m_i') \in \mathcal{M} \times \mathcal{M} \text{ s.t. } \mathbf{X}(m_i) \ominus \mathbf{X}(m_i') = \boldsymbol{\xi}_i\big].$$

Consequently, the event $\big[\{\boldsymbol{\xi}_1,\ldots,\boldsymbol{\xi}_\ell\} \subseteq (\mathcal{C}\ominus\mathcal{C})^\star\big]$ is equivalent to the event

$$\Big[\exists\{(m_i, m_i')\}_{i=1}^\ell \subseteq \mathcal{M} \times \mathcal{M} \ \text{ s.t. } \mathbf{X}(m_i) \ominus \mathbf{X}(m_i') = \boldsymbol{\xi}_i,\ \forall i \in [\ell]\Big]. \tag{C.18}$$

Hence, by the Union Bound,

$$\mathbb{P}\big[\{\boldsymbol{\xi}_1,\ldots,\boldsymbol{\xi}_\ell\} \subseteq (\mathcal{C}\ominus\mathcal{C})^\star\big]$$
$$\leq \sum_{\{(m_i,m_i')\}_{i=1}^\ell \subseteq \mathcal{M}\times\mathcal{M}} \mathbb{P}\bigcap_{i\in[\ell]} \big[\mathbf{X}(m_i) \ominus \mathbf{X}(m_i') = \boldsymbol{\xi}_i\big] \tag{C.19}$$
$$\leq 2^{2n\mathrm{R}\ell}|\mathcal{A}|^{-n\ell}, \tag{C.20}$$

where (C.20) follows from Lemma C.3 ahead and from the fact that the number of terms on the RHS of (C.19) is upper bounded by $2^{2n\mathrm{R}\ell}$.

$\square$

From (C.13) and (C.16) we now obtain

$$\mathbb{P}\big[\max_{\mathbf{z}\in\mathcal{S}^n} |\mathcal{F}_\mathbf{z}| > \mathsf{L} - 1\big]$$
$$\leq \sum_{\mathbf{z}\in\mathcal{S}^n} \mathbb{P}\big[\exists \text{ tri-independent } \boldsymbol{\xi}_1,\ldots,\boldsymbol{\xi}_\ell \in \mathcal{S}^n \ominus \mathbf{z}$$
$$\text{s.t. } \{\boldsymbol{\xi}_1,\ldots,\boldsymbol{\xi}_\ell\} \subseteq (\mathcal{C}\ominus\mathcal{C})^\star\big] \tag{C.21}$$

$$\leq \sum_{\substack{\mathbf{z} \in \mathcal{S}^n \text{ tri-independent} \\ \boldsymbol{\xi}_1, \ldots, \boldsymbol{\xi}_\ell \in \mathcal{S}^n \ominus \mathbf{z}}} \mathbb{P}\big[\{\boldsymbol{\xi}_1, \ldots, \boldsymbol{\xi}_\ell\} \subseteq (\mathcal{C} \ominus \mathcal{C})^\star\big] \tag{C.22}$$

$$\leq |\mathcal{S}|^n \binom{|\mathcal{S}|^n}{\ell} \left(\frac{2^{2n\mathsf{R}}}{|\mathcal{A}|^n}\right)^\ell \tag{C.23}$$

$$\leq \frac{|\mathcal{S}|^{n(\ell+1)}}{\ell!} \left(\frac{2^{2n\mathsf{R}}}{|\mathcal{A}|^n}\right)^\ell, \tag{C.24}$$

where (C.22) follows from the Union Bound, and (C.23) follows from Claim C.2.

From (C.9) and (C.24) (and the Union Bound) we infer that, the two properties (i) and (ii) hold simultaneously with probability strictly larger than

$$1 - \frac{2^{2n\mathsf{R}}}{2|\mathcal{A}|^n} - \frac{|\mathcal{S}|^{n(\ell+1)}}{\ell!}\left(\frac{2^{2n\mathsf{R}}}{|\mathcal{A}|^n}\right)^\ell \geq 1 - \frac{1}{2}|\mathcal{S}|^{-n\frac{\ell+1}{\ell}} - \frac{1}{\ell!} \tag{C.25}$$

$$\geq 1 - \frac{1}{2} - \frac{1}{\ell!} \tag{C.26}$$

$$\geq 0, \tag{C.27}$$

where (C.25) holds because, by (5.74) and (C.6)

$$2^{2n\mathsf{R}} \leq 2^{2n\left(\frac{1}{2}\log\frac{|\mathcal{A}|}{|\mathcal{S}|} - \frac{\log|\mathcal{S}|}{2\ell}\right)} = |\mathcal{A}|^n |\mathcal{S}|^{-n\frac{\ell+1}{\ell}}. \tag{C.28}$$

Thus, with positive probability, the random codebook $\mathcal{C}$ satisfies both desired properties simultaneously. This concludes the proof of Lemma 5.20 (assuming Lemma C.3 ahead). $\qquad\square$

C.3 Lemma C.3 and Its Proof

We next state and prove Lemma C.3.

Lemma C.3. *Let $\{(m_i, m_i')\}_{i=1}^\ell \subseteq \mathcal{M} \times \mathcal{M}$ be ℓ message tuples. Let the undirected graph $G = (V, E)$ be of vertices $V = \mathcal{M}$ and of edges $E = \{e_i\}_{i=1}^\ell$, where e_i denotes the edge $\{m_i, m_i'\}$. Let the ℓ n-tuples $\{\boldsymbol{\xi}_i\}_{i=1}^\ell$ be tri-independent. If $\mathbf{X}(1), \ldots, \mathbf{X}(|\mathcal{M}|)$ are drawn IID, each equiprobably from $\mathcal{A}^n$, then the probability*

$$\mathbb{P}\bigcap_{i \in [\ell]} \big[\mathbf{X}(m_i) \ominus \mathbf{X}(m_i') = \boldsymbol{\xi}_i\big] \tag{C.29}$$

equals $|\mathcal{A}|^{-n\ell}$ if the graph G is acyclic[1], and equals zero otherwise.

Proof. First, assume that G contains some cycle, say of length k, containing vertices v_1, v_2, ..., v_k, $v_{k+1} \triangleq v_1$. By possibly permuting $\{\boldsymbol{\xi}_i, (m_i, m_i')\}_{i=1}^{\ell}$ we may assume without loss of generality that $\{v_1, v_2\} = \{m_1, m_1'\}$ (as sets), so either $(v_1, v_2) = (m_1, m_1')$ or $(v_1, v_2) = (m_1', m_1)$ (or both, if this is a loop). In the former case the event of interest is $\left[\mathbf{X}(m_1) \ominus \mathbf{X}(m_1') = \boldsymbol{\xi}_1\right]$ and in the latter $\left[\mathbf{X}(m_1') \ominus \mathbf{X}(m_1) = \boldsymbol{\xi}_1\right]$. We similarly assume that $\{v_i, v_{i+1}\} = \{m_i, m_{i+1}'\}$ (as sets) for all $i \in [k]$.

The probability (C.29) is positive if, and only if, there exist realizations $\mathbf{x}_1, \ldots, \mathbf{x}_{|\mathcal{M}|} \in \mathcal{A}^n$ such that

$$\mathbf{x}_{m_i} \ominus \mathbf{x}_{m_i'} = \boldsymbol{\xi}_i, \quad i \in [\ell], \tag{C.30}$$

which, as we shall see, contradicts the nondegeneracy of the n-tuples.

Define

$$\epsilon_i = \begin{cases} 1 & \text{if } (v_i, v_{i+1}) = (m_i, m_i'), \\ -1 & \text{otherwise,} \end{cases} \quad i \in [k]. \tag{C.31}$$

Then, (C.30) and (C.31) imply

$$\sum_{i=1}^{k} \epsilon_i \boldsymbol{\xi}_i = \sum_{i=1}^{k} \epsilon_i \left(\mathbf{x}_{m_i} \ominus \mathbf{x}_{m_i'}\right) \tag{C.32}$$

$$= \sum_{i=1}^{k} \left(\mathbf{x}_{v_i} \ominus \mathbf{x}_{v_{i+1}}\right) \tag{C.33}$$

$$= \mathbf{0}, \tag{C.34}$$

where the last equality holds because $v_{k+1} = v_1$.

By setting $\epsilon_i = 0$ for all $i \in [\ell] \setminus [k]$ (i.e., for all edges not in the cycle), we obtain ℓ coefficients $\{\epsilon_i\}_{i=1}^{\ell}$ (that are not all zero) taking values in $\{0, \pm 1\}$ such that $\sum_{j=1}^{\ell} \epsilon_j \boldsymbol{\xi}_j = \sum_{j=1}^{k} \epsilon_j \boldsymbol{\xi}_j = \mathbf{0}$, which contradicts the tri-independence assumption.

We next consider the case where the graph G is acyclic. *A fortiori*, it contains no loops, so $m_i \neq m_i'$ for all $i \in [\ell]$. Proving that

$$\mathbb{P} \bigcap_{i \in [\ell]} \left[\mathbf{X}(m_i) \ominus \mathbf{X}(m_i') = \boldsymbol{\xi}_i\right] = \frac{1}{|\mathcal{A}|^{n\ell}} \tag{C.35}$$

---

[1]G is allowed to contain loops and parallel edges, which also count as cycles.

is tantamount to proving that the events $\left\{\left[\mathbf{X}(m_i) \ominus \mathbf{X}(m_i') = \boldsymbol{\xi}_i\right]\right\}_{i \in [\ell]}$ are independent, because it can be readily verified that for any $m \neq m'$ and $\boldsymbol{\xi} \in \mathcal{A}^n$, $\mathbb{P}\left[\mathbf{X}(m) \ominus \mathbf{X}(m') = \boldsymbol{\xi}\right] = \frac{1}{|\mathcal{A}|^n}$.

Because the codewords are chosen independently, the events corresponding to edges in different connected components of G are independent. We can therefore focus on an arbitrary non-empty connected component, denoted $G_1 = (V_1, E_1)$, and show that the events corresponding to E_1 are independent. That is, we need to show that

$$\mathbb{P} \bigcap_{i:\, e_i \in E_1} \left[\mathbf{X}(m_i) \ominus \mathbf{X}(m_i') = \boldsymbol{\xi}_i\right] = \frac{1}{|\mathcal{A}|^{n|E_1|}}. \tag{C.36}$$

To prove this, we will show by induction on $|E_1|$ that the system of linear equations (with $|E_1|$ equations and $|V_1|$ variables) corresponding to this connected component, namely,

$$\mathbf{x}_{m_i} \ominus \mathbf{x}_{m_i'} = \boldsymbol{\xi}_i, \qquad i \in \{j:\, e_j \in E_1\} \tag{C.37}$$

has $|\mathcal{A}|^n$ solutions. This is to be expected because G is acyclic and connected, so G_1 is a tree, and hence $|V_1| = |E_1| + 1$.

If $|E_1| = 1$, there are two variables and one equation, so the number of solutions is, indeed, $|\mathcal{A}|^n$. If $|E_1| \geq 2$, let m_0 be a degree one vertex of G_1 (which is guaranteed to exist because G_1 is a tree). As such, $\mathbf{x}(m_0)$ appears in only one of the equations, and it is therefore uniquely determined by the remaining $(|V_1| - 1)$ variables. The number of solutions is thus as for the system that remains when we remove $\mathbf{x}_{m_0}$ and the equation in which it appears from the system, which leaves us with $(|E_1| - 1)$ equations corresponding to the induced subgraph $G_1[V_1 \setminus \{m_0\}]$. This subgraph is still a tree, and hence, by the induction hypothesis, this restricted system with $(|E_1| - 1)$ equations and $(|V_1| - 1)$ variables has $|\mathcal{A}|^n$ solutions. This concludes the induction and establishes that, as we claimed, the system of equations (C.37) has $|\mathcal{A}|^n$ solutions.

We can now complete the proof of (C.36):

$$\mathbb{P} \bigcap_{i:\, e_i \in E_1} \left[\mathbf{X}(m_i) \ominus \mathbf{X}(m_i') = \boldsymbol{\xi}_i\right]$$

$$= \sum_{\substack{\{\mathbf{x}_m\}_{m \in V_1} \\ \text{solving (C.37)}}} \mathbb{P} \bigcap_{m \in V_1} \left[\mathbf{X}(m) = \mathbf{x}_m\right] \tag{C.38}$$

$$= \sum_{\substack{\{\mathbf{x}_m\}_{m \in V_1} \\ \text{solving (C.37)}}} \frac{1}{|\mathcal{A}|^{n|V_1|}} \tag{C.39}$$

$$= \frac{|\mathcal{A}|^n}{|\mathcal{A}|^{n|V_1|}} \tag{C.40}$$

$$= \frac{1}{|\mathcal{A}|^{n|E_1|}}, \tag{C.41}$$

where the last equality holds because G_1 is a tree, so $|V_1| = |E_1| + 1$. $\qquad\square$

Appendix D

Proofs for Chapter 6

D.1   Proof of Proposition 6.3

To see why the maximum in (6.7) is unaltered when $Q_{X|U,V}$ is restricted to be deterministic, we fix $Q_S, Q_V, Q_{U|S,V}, Q_{Y|X,S}$ and note that this fixes $I(U;S|V)$ and causes $I(U;Y|V)$ to be a function of $Q_{X|U,V}$ only. Since $I(U;Y|V)$ is convex in $Q_{Y|U,V}$, and since the latter is linear in $Q_{X|U,V}$, it follows that $I(U;Y|V)$ is convex in $Q_{X|U,V}$, and thus establishes that the maximization can be achieved with $Q_{X|U,V}$ being deterministic.

To derive cardinality bounds on $\mathcal{V}$ and $\mathcal{U}$, we first rewrite (6.7) as

$$
\mathsf{C}^{(\mathrm{I})}(\mathsf{R_h}) = \max_{\substack{Q_V, \{\mathsf{R}_{h,v}\}_{v \in \mathcal{V}} : \\ \mathbb{E}[\mathsf{R}_{h,V}] = \mathsf{R_h}}} \sum_{v \in \mathcal{V}} Q_V(v) \left(\max_{\substack{Q_{U|S}, Q_{X|U} : \\ I(U;S) \leq \mathsf{R}_{h,v}}} I(U;Y) - I(U;S) + \mathsf{R}_{h,v} \right).
$$

$$\tag{D.1}$$

To find an upper bound on $\mathcal{U}$, we fix $\mathsf{R}_{h,v}$ and focus on the parenthesized term in (D.1). We follow the same technique as in [58, Appendix C]. Fix any $U \in \mathcal{U}$ such that $S \; \multimap \; U \; \multimap \; X$. Consider the set $\mathcal{P}_\mathsf{X}$ of all product PMFs on $\mathcal{X} \times \mathcal{S}$, which is connected and compact, and, by the Markov relation, contains $Q_{X,S|U=u}$ for all $u \in \mathcal{U}$. The following $(|\mathcal{X}| \cdot |\mathcal{S}| + 1)$ functions of $\pi \in \mathcal{P}_\mathsf{X}$ are all continuous:

$$g_1(\pi) = H(Y), \tag{D.2}$$
$$g_2(\pi) = H(S), \tag{D.3}$$

$$g_{i,j}(\pi) = \pi(i,j), \quad (i,j) \in \mathcal{X} \times \mathcal{S} \setminus \{(|\mathcal{X}|, |\mathcal{S}|)\}, \tag{D.4}$$

where the continuity of $g_1(\cdot)$ and $g_2(\cdot)$ follows from the continuity of entropy and the linearity of Q_Y in π. Applying the Support Lemma [58, Appendix C], we can find a random variable U', taking at most $(|\mathcal{X}| \cdot |\mathcal{S}| + 1)$ values in $\mathcal{U}' \subseteq \mathcal{U}$, such that, when substituting U' for U, the values of $H(Y|U')$ and $H(S|U')$ and the distribution $Q_{X,S}$ are preserved. The latter implies that Q_S and Q_Y are preserved, and hence also $H(Y)$ and $H(S)$. To sum up, it is therefore sufficient to restrict $|\mathcal{U}| \leq |\mathcal{X}| \cdot |\mathcal{S}| + 1$, while preserving the values of Q_S, $I(U;Y)$, and $I(U;S)$, and preserving the Markov relation $S \multimap U \multimap X$.

To prove the desired upper bound on $\mathcal{V}$, let $g(\mathsf{R}_{\mathrm{h},v})$ denote the term in the parenthesis of (D.1). Given any choice of $\mathcal{V}$, Q_V, and $\{\mathsf{R}_{\mathrm{h},v}\}_{v \in \mathcal{V}}$, we write the two-dimensional vector $\left(\mathsf{R}_{\mathrm{h}}, \sum_{v \in \mathcal{V}} Q_V(v) g(\mathsf{R}_{\mathrm{h},v})\right)$ as the convex combination

$$\begin{pmatrix} \mathsf{R}_{\mathrm{h}} \\ \sum_{v \in \mathcal{V}} Q_V(v) g(\mathsf{R}_{\mathrm{h},v}) \end{pmatrix} = \sum_{v \in \mathcal{V}} Q_V(v) \begin{pmatrix} \mathsf{R}_{\mathrm{h},v} \\ g(\mathsf{R}_{\mathrm{h},v}) \end{pmatrix}. \tag{D.5}$$

Applying Carathéodory's Theorem [58, Appendix A], we obtain that it is sufficient to have $|\mathcal{V}| \leq 3$.

Indeed, $\mathcal{V}$ can be further restricted to have $|\mathcal{V}| \leq 2$. To see the strengthened upper bound, we will show that the region

$$\{(r, g(r)) \colon r \in \mathbb{R}\} \subset \mathbb{R}^2 \tag{D.6}$$

is connected and compact and apply the Fenchel–Eggleston–Carathéodory Theorem [58, Appendix A]. This is tantamount to proving the continuity of the function $g(\cdot)$, which can be observed using the Berge's Maximum Theorem [64, Chapter 6] by noting that $I(U;S)$ is convex in $Q_{U|S}$ and that all the mutual information is continuous in the joint distribution.

$\square$

D.2 Proof of Theorem 6.12

In light of (6.150), we only need to establish the achievability of $\mathsf{C}_\ell^{(\rho)}(\mathsf{R}_{\mathrm{h}})$.[1]

---

[1] This will also imply the achievability of $\mathsf{C}(\mathsf{R}_{\mathrm{h}})$, rendering Section 6.5.1 superfluous. Nevertheless, we keep Section 6.5.1 because the proof there is simpler and lays the foundations for the one here.

Our coding scheme is similar to the one of Section 6.5.1.2, but with some additional elements from Section 4.3, e.g., that the helper also describes the normalized squared-norm $\|Z^n\|^2/n$ of the noise sequence. To that end, it uses a quantizer whose precision increases with the blocklength but whose number of levels is sub-exponential in the blocklength so that its output is describable with zero rate. This description allows the encoder to employ appropriate scaling so that the ratio between the energy in the transmitted sequence and the energy in the noise sequence lies, with high probability, in an interval (near the SNR A) that shrinks with the blocklength. This allows us to control the zero-error list even on the rare occasions when the noise is large.

More specifically, consider the interval $[0, \mathsf{M}]$, where the constant M is chosen large enough so that the large-deviation probability of overload $\mathbb{P}[\|Z^n\|^2 \geq n\mathsf{M}]$ decays sufficiently fast in n to guarantee that—even if an overload results in the list containing all $2^{n\mathsf{R}}$ codewords—the contribution of the overload to the ρ-th moment of the list be negligible:

$$\lim_{n \to \infty} 2^{n\rho\mathsf{R}} \cdot \mathbb{P}\big[\|Z^n\|^2 \geq n\mathsf{M}\big] = 0. \tag{D.7}$$

As the analysis of (4.161) in Section 4.3.4.1, we can now choose $\mathsf{M} = \max\{2\mathsf{N},\ 7\mathsf{N}\rho\mathsf{C}(\mathsf{R_h})\}$.

Fix ϵ' and ϵ (later to tend to zero) with $0 < \epsilon' < \epsilon < \min\{\mathsf{R_h}, \mathsf{P}\}$. For given positive integer K (which determines the number of quantization cells), consider the partition of the interval $[0, \mathsf{M}]$ into K subintervals, each of length $\Delta \triangleq \mathsf{M}/\mathsf{K}$. The choice of K depends on M and ϵ and will be specified later in (D.17).

Let $\theta_0 \in [0, \pi/2]$ be such that

$$\sin\theta_0 = 2^{-(\mathsf{R_h}-\epsilon)}. \tag{D.8}$$

The codebook we use is a slightly shrunk and sparsified version of the one in Section 6.5.1.2: Let $\mathcal{C}$ be a subset of $\partial\mathcal{B}\big(\sqrt{n(\mathsf{P} - \epsilon)}\big)$ with $2^{n(\mathsf{R_h}-\epsilon')}$ codewords that completely cover the sphere with caps of half-angle θ_0. (The existence of such a code for every sufficiently large n can be inferred from [59].) Generate $2^{n\mathsf{R}}$ independent random orthogonal transformations indexed by the messages, and for each message $m \in \mathcal{M}$, let $\mathcal{C}_m = \{x^n(m,t)\}_{t\in\{1,\dots,2^{n(\mathsf{R_h}-\epsilon')}\}}$ be the result of applying the m-th random orthogonal transformation to the elements of $\mathcal{C}$.

The helper, upon observing the noise sequence Z^n and the message $m \in \mathcal{M}$, produces two descriptions $T_1 \in \{0,\dots,\mathsf{K}\}$ and $T_2 \in$

$\{0, \ldots, 2^{n(\mathsf{R_h}-\epsilon')}\}$, where T_1 describes the energy of the noise sequence

$$T_1 = \phi_1^{(n)}(Z^n) = \begin{cases} \left\lceil \frac{\|Z^n\|^2}{n\Delta} \right\rceil & \text{if } \|Z^n\|^2 < n\mathsf{M}, \\ 0 & \text{otherwise,} \end{cases} \tag{D.9}$$

and $T_2 = \phi_2^{(n)}(m, Z^n)$ is, as in Section 6.5.1.2, the index of a codeword in $\mathcal{C}_m$ whose angle with Z^n does not exceed θ_0

$$\angle\big(x^n(m, T_2), Z^n\big) \le \theta_0. \tag{D.10}$$

The rate of help is therefore upper-bounded by

$$\frac{1}{n} \log \big(\mathsf{K} 2^{n(\mathsf{R_h}-\epsilon')}\big) \tag{D.11}$$

which, for sufficiently large n, does not exceed $\mathsf{R_h}$.

To convey the m-th message, the encoder transmits

$$U^n(m, T_1, T_2) = \begin{cases} x^n(m, T_2) \cdot \sqrt{\frac{\Delta T_1}{\mathsf{N}}} & \text{if } T_1 \ne 0 \\ 0^n & \text{if } T_1 = 0 \end{cases} \tag{D.12}$$

i.e., a scaled version of the codeword if the quantization is successful, and the all-zero sequence otherwise. The expected energy in the transmitted sequence is therefore upper-bounded by

$$\mathbb{E}_{T_1, T_2}\left[\sum_{k=1}^{n} U_k(m, T_1, T_2)^2 \right] = \mathbb{E}_{T_1}\left[n(\mathsf{P} - \epsilon) \cdot \frac{\Delta T_1}{\mathsf{N}} \right] \tag{D.13}$$

$$\le \mathbb{E}_{Z^n}\left[n(\mathsf{P} - \epsilon) \cdot \frac{\Delta}{\mathsf{N}} \left(\frac{\|Z^n\|^2}{n\Delta} + 1 \right) \right] \tag{D.14}$$

$$= n(\mathsf{P} - \epsilon)\left(1 + \frac{\Delta}{\mathsf{N}} \right) \tag{D.15}$$

$$\le n\mathsf{P}, \tag{D.16}$$

where the last inequality holds whenever K is chosen such that M/K, namely Δ, satisfies

$$\frac{\mathsf{M}}{\mathsf{K}} \le \frac{\mathsf{N}\epsilon}{\mathsf{P} - \epsilon}. \tag{D.17}$$

The decoder—based on its observation Y^n and the help (T_1, T_2)—produces the "remotely-plausible list"

$$\mathcal{L}_0(Y^n, T_1, T_2) = \Big\{ m' \in \mathcal{M} : \exists \xi^n \text{ s.t. } \phi_1^{(n)}(\xi^n) = T_1,$$

$$\phi_2^{(n)}(m', \xi^n) = T_2,$$

$$\text{and } U^n(m', T_1, T_2) + \xi^n = Y^n \Big\}. \tag{D.18}$$

To analyze the ρ-th moment of $\big| \mathcal{L}_0(Y^n, T_1, T_2) \big|$, we first utilize the Law of Total Expectation

$$\mathbb{E}\big[\big| \mathcal{L}_0(Y^n, T_1, T_2) \big|^\rho \big]$$
$$= \mathbb{E}\big[\big| \mathcal{L}_0(Y^n, T_1, T_2) \big|^\rho \,\big|\, T_1 \neq 0 \big] \mathbb{P}\big[T_1 \neq 0 \big]$$
$$\qquad + \mathbb{E}\big[\big| \mathcal{L}_0(Y^n, T_1, T_2) \big|^\rho \,\big|\, T_1 = 0 \big] \mathbb{P}\big[T_1 = 0 \big] \tag{D.19}$$
$$\leq \mathbb{E}\big[\big| \mathcal{L}_0(Y^n, T_1, T_2) \big|^\rho \,\big|\, T_1 \neq 0 \big] \mathbb{P}\big[T_1 \neq 0 \big] + 2^{n\rho R} \mathbb{P}\big[T_1 = 0 \big], \tag{D.20}$$

where the second term on the RHS of (D.20) tends to zero as n tends to infinity by (D.7).

To upper-bound the first term on the RHS of (D.20), we first prove that, when the quantization is successful, i.e., when $T_1 \neq 0$, for any $m' \in \mathcal{L}_0(Y^n, T_1, T_2)$, the angle between the codeword $x^n(m', T_2)$ and Y^n is small. Indeed, for any m' in the list, there exists ξ^n such that

$$\| \xi^n \|^2 \leq n \Delta T_1, \tag{D.21}$$

and the angle between ξ^n and $U^n(m', T_1, T_2)$, which we denote by θ, satisfies

$$\theta = \angle(U^n(m', T_1, T_2), \xi^n) \tag{D.22}$$
$$= \angle(x^n(m', T_2), \xi^n) \tag{D.23}$$
$$\leq \theta_0, \tag{D.24}$$

where (D.23) holds because $U^n(m', T_1, T_2)$ is a scaled version of $x^n(m, T_2)$. Arguing as we did in Section 6.5.1.2, we conclude that the angle between U^n and Y^n, which we denote by α, satisfies

$$\sin \alpha = \left(\sqrt{\frac{\|U^n\|^2}{\|\xi^n\|^2} + 1 + 2\frac{\|U^n\|}{\|\xi^n\|} \cos\theta} \right)^{-1} \sin\theta \tag{D.25}$$

$$\leq \left(\sqrt{\frac{\mathsf{P} - \epsilon}{\mathsf{N}} + 1 + 2\sqrt{\frac{\mathsf{P} - \epsilon}{\mathsf{N}}} \cos \theta_0} \right)^{-1} \sin \theta_0 \qquad \text{(D.26)}$$

$$\triangleq \sin \alpha_0, \qquad \text{(D.27)}$$

where (D.27) holds because $\theta \leq \theta_0$ and

$$\frac{\|U^n\|^2}{\|\xi^n\|^2} = n(\mathsf{P} - \epsilon) \cdot \frac{\Delta T_1}{\mathsf{N}} \cdot \frac{1}{\|\xi^n\|^2} \geq \frac{\mathsf{P} - \epsilon}{\mathsf{N}}. \qquad \text{(D.28)}$$

Therefore, $\mathcal{L}_0(Y^n, T_1, T_2)$ is a subset of the "close-to-output list"

$$\tilde{\mathcal{L}}(Y^n, T_2) \triangleq \{ m' \in \mathcal{M} \colon \angle(x^n(m', T_2), Y^n) \leq \alpha_0 \}, \qquad \text{(D.29)}$$

and it is sufficient to prove that

$$\lim_{n \to \infty} \mathbb{E}\left[|\tilde{\mathcal{L}}(Y^n, T_2)|^\rho \,\big|\, T_1 \neq 0 \right] = 1. \qquad \text{(D.30)}$$

Indeed, conditional on any $T_1 \neq 0$, any transmitted message M, and any received tuple (Y^n, T_2), the distribution of $|\tilde{\mathcal{L}}(Y^n, T_2)|$ is that of

$$1 + \mathrm{Bin}\left(|\mathcal{M}| - 1, \frac{C_n(\alpha_0)}{C_n(\pi)} \right) = 1 + \mathrm{Bin}\left(2^{n\mathsf{R}} - 1, \, 2^{n(\log \sin \alpha_0 + o(1))} \right) \qquad \text{(D.31)}$$

where $\mathrm{Bin}(n, p)$ denotes a binomial random variable with n trials and success probability p. By [23, Lemma 1.1], its ρ-th moment tends to one whenever

$$\mathsf{R} < -\log \sin \alpha_0 \qquad \text{(D.32)}$$

$$= -\log \left(\left(\sqrt{\frac{\mathsf{P} - \epsilon}{\mathsf{N}} + 1 + 2\sqrt{\frac{\mathsf{P} - \epsilon}{\mathsf{N}}} \cos \theta_0} \right)^{-1} \sin \theta_0 \right) \qquad \text{(D.33)}$$

$$= \frac{1}{2} \log \left(\frac{\mathsf{P} - \epsilon}{\mathsf{N}} + 1 + 2\sqrt{\frac{\mathsf{P} - \epsilon}{\mathsf{N}}} \sqrt{1 - 2^{-2(\mathsf{R}_\mathrm{h} - \epsilon)}} \right) + \mathsf{R}_\mathrm{h} - \epsilon. \qquad \text{(D.34)}$$

The direct part now follows upon letting ϵ tend to zero. $\qquad \square$

Bibliography

[1] S. I. Bross and A. Lapidoth, "The Gaussian state-dependent channel with rate-limited decoder state-information," in *2016 IEEE International Conference on the Science of Electrical Engineering (ICSEE)*, 2016, pp. 1–5.

[2] S. I. Bross, A. Lapidoth, and G. Marti, "Decoder-assisted communications over additive noise channels," *IEEE Transactions on Communications*, vol. 68, no. 7, pp. 4150–4161, 2020.

[3] A. Lapidoth and G. Marti, "Encoder-assisted communications over additive noise channels," *IEEE Transactions on Information Theory*, vol. 66, no. 11, pp. 6607–6616, 2020.

[4] S. I. Bross and A. Lapidoth, "The additive noise channel with a helper," in *2019 IEEE Information Theory Workshop (ITW)*, 2019, pp. 1–5.

[5] N. Merhav, "On error exponents of encoder-assisted communication systems," *IEEE Transactions on Information Theory*, vol. 67, no. 11, pp. 7019–7029, 2021.

[6] S. Loyka and N. Merhav, "The secrecy capacity of the Gaussian wiretap channel with rate-limited help at the decoder," in *2022 IEEE International Symposium on Information Theory (ISIT)*, 2022, pp. 1040–1045.

[7] A. Lapidoth and B. Ni, "Assisted identification over modulo-additive noise channels," *Entropy*, vol. 25, no. 9, 2023. [Online]. Available: https://www.mdpi.com/1099-4300/25/9/1314

[8] A. Khina and N. Merhav, "Modulation and estimation with a helper," *IEEE Transactions on Information Theory*, vol. 70, no. 9, pp. 6189–6210, 2024.

[9] C. E. Shannon, "Channels with side information at the transmitter," *IBM Journal of Research and Development*, vol. 2, no. 4, pp. 289–293, 1958.

[10] S. I. Gel'fand and M. S. Pinsker, "Coding for channels with random parameters," *Probl. Control Inf. Theory*, vol. 9, no. 1, pp. 19–31, 1980.

[11] C. Heegard and A. El Gamal, "On the capacity of computer memory with defects," *IEEE Transactions on Information Theory*, vol. 29, no. 5, pp. 731–739, 1983.

[12] A. Rosenzweig, Y. Steinberg, and S. Shamai, "On channels with partial channel state information at the transmitter," *IEEE Transactions on Information Theory*, vol. 51, no. 5, pp. 1817–1830, 2005.

[13] Y. Steinberg, "Coding for channels with rate-limited side information at the decoder, with applications," *IEEE Transactions on Information Theory*, vol. 54, no. 9, pp. 4283–4295, 2008.

[14] G. Keshet, Y. Steinberg, and N. Merhav, "Channel coding in the presence of side information," *Foundations and Trends® in Communications and Information Theory*, vol. 4, no. 6, pp. 445–586, 2008. [Online]. Available: http://dx.doi.org/10.1561/0100000025

[15] A. Lapidoth and L. Wang, "State-dependent DMC with a causal helper," *IEEE Transactions on Information Theory*, vol. 70, no. 5, pp. 3162–3174, 2024.

[16] A. Lapidoth and Y. Steinberg, "The state-dependent channel with a rate-limited cribbing helper," *Entropy*, vol. 26, no. 7, 2024. [Online]. Available: https://www.mdpi.com/1099-4300/26/7/570

[17] C. E. Shannon, "A mathematical theory of communication," *The Bell System Technical Journal*, vol. 27, no. 3, pp. 379–423, 1948.

[18] M. S. Pinsker and A. Y. Sheverdjaev, "Transmission capacity with zero error and erasure," *Problemy Peredachi Informatsii*, vol. 6, no. 1, pp. 20–24, 1970.

[19] I. Csiszár and P. Narayan, "Channel capacity for a given decoding metric," *IEEE Transactions on Information Theory*, vol. 41, no. 1, pp. 35–43, 1995.

[20] C. Bunte, A. Lapidoth, and A. Samorodnitsky, "The zero-undetected-error capacity approaches the Sperner capacity," *IEEE Transactions on Information Theory*, vol. 60, no. 7, pp. 3825–3833, 2014.

[21] R. Ahlswede, N. Cai, and Z. Zhang, "Erasure, list, and detection zero-error capacities for low noise and a relation to identification," *IEEE Transactions on Information Theory*, vol. 42, no. 1, pp. 55–62, 1996.

[22] E. Arıkan, "An inequality on guessing and its application to sequential decoding," *IEEE Transactions on Information Theory*, vol. 42, no. 1, pp. 99–105, 1996.

[23] C. Bunte and A. Lapidoth, "On the listsize capacity with feedback," *IEEE Transactions on Information Theory*, vol. 60, no. 11, pp. 6733–6748, 2014.

[24] İ. E. Telatar, "Zero-error list capacities of discrete memoryless channels," *IEEE Transactions on Information Theory*, vol. 43, no. 6, pp. 1977–1982, 1997.

[25] T. M. Cover and J. A. Thomas, *Elements of information theory*, 2nd ed.   John Wiley & Sons, Ltd, 2005.

[26] S. M. Moser, *Advanced topics in information theory (lecture notes)*, 5th ed., 2022, https://moser-isi.ethz.ch/docs/atit_script_v58.pdf.

[27] ——, *Information theory (lecture notes)*, 6th ed., 2018, https://moser-isi.ethz.ch/docs/it_script_v616.pdf.

[28] Y. Polyanskiy and Y. Wu, *Information Theory: From Coding to Learning*.   Cambridge University Press, 2023+.

[29] İ. E. Telatar, "Multi-access communications with decision feedback decoding," Ph.D. dissertation, Massachusetts Institute of Technology, 1992.

[30] İ. E. Telatar and R. G. Gallager, "New exponential upper bounds to error and erasure probabilities," in *Proceedings of 1994 IEEE International Symposium on Information Theory*, 1994, pp. 379–.

[31] C. Bunte, A. Lapidoth, and A. Samorodnitsky, "The zero-undetected-error capacity of the low-noise cyclic triangle channel," in *2013 IEEE International Symposium on Information Theory*, 2013, pp. 91–95.

[32] C. Bunte and A. Lapidoth, "The zero-undetected-error capacity of discrete memoryless channels with feedback," in *2012 50th Annual Allerton Conference on Communication, Control, and Computing (Allerton)*, 2012, pp. 1838–1842.

[33] A. Ganti, A. Lapidoth, and İ. E. Telatar, "Mismatched decoding revisited: General alphabets, channels with memory, and the wide-band limit," *IEEE Transactions on Information Theory*, vol. 46, no. 7, pp. 2315–2328, 2000.

[34] E. Arıkan, "Channel combining and splitting for cutoff rate improvement," *IEEE Transactions on Information Theory*, vol. 52, no. 2, pp. 628–639, 2006.

[35] ——, "Channel polarization: A method for constructing capacity-achieving codes for symmetric binary-input memoryless channels," *IEEE Transactions on Information Theory*, vol. 55, no. 7, pp. 3051–3073, 2009.

[36] ——, "On the origin of polar coding," *IEEE Journal on Selected Areas in Communications*, vol. 34, no. 2, pp. 209–223, 2016.

[37] R. G. Gallager, *Information theory and reliable communication.* USA: John Wiley & Sons, Inc., 1968.

[38] ——, "A simple derivation of the coding theorem and some applications," *IEEE Transactions on Information Theory*, vol. 11, no. 1, pp. 3–18, 1965.

[39] I. Csiszár, "Generalized cutoff rates and Rényi's information measures," *IEEE Transactions on Information Theory*, vol. 41, no. 1, pp. 26–34, 1995.

[40] J. Massey, "Guessing and entropy," in *Proceedings of 1994 IEEE International Symposium on Information Theory*, 1994, pp. 204–.

[41] C. Bunte and A. Lapidoth, "Encoding tasks and Rényi entropy," *IEEE Transactions on Information Theory*, vol. 60, no. 9, pp. 5065–5076, 2014.

[42] S. Fehr and S. Berens, "On the conditional Rényi entropy," *IEEE Transactions on Information Theory*, vol. 60, no. 11, pp. 6801–6810, 2014.

[43] S. Verdú, "Error exponents and α-mutual information," *Entropy*, vol. 23, no. 2, 2021. [Online]. Available: https://www.mdpi.com/1099-4300/23/2/199

[44] A. Lapidoth and N. Miliou, "Duality bounds on the cutoff rate with applications to Ricean fading," *IEEE Transactions on Information Theory*, vol. 52, no. 7, pp. 3003–3018, 2006.

[45] C. Pfister, "On Rényi information measures and their applications," Ph.D. dissertation, ETH Zurich, 2019.

[46] H. P. Rosenthal, "On the subspaces of L^p $(p > 2)$ spanned by sequences of independent random variables," *Israel Journal of Mathematics*, vol. 8, no. 3, pp. 273–303, 1970.

[47] C. E. Shannon, "The zero error capacity of a noisy channel," *IRE Transactions on Information Theory*, vol. 2, no. 3, pp. 8–19, 1956.

[48] L. Lovász, "On the Shannon capacity of a graph," *IEEE Transactions on Information Theory*, vol. 25, no. 1, pp. 1–7, 1979.

[49] G. Marti, "Channels with a helper," in *Master's thesis, ETH Zurich*, 2019.

[50] A. Bracher and A. Lapidoth, "The zero-error feedback capacity of state-dependent channels," *IEEE Transactions on Information Theory*, vol. 64, no. 5, pp. 3538–3578, 2018.

[51] P. Elias, "Zero error capacity under list decoding," *IEEE Transactions on Information Theory*, vol. 34, no. 5, pp. 1070–1074, 1988.

[52] V. V. Zyablov and M. S. Pinsker, "List concatenated decoding," *Problemy Peredachi Informatsii*, vol. 17, no. 4, pp. 29–33, 1981.

[53] P. Elias, "Error-correcting codes for list decoding," *IEEE Transactions on Information Theory*, vol. 37, no. 1, pp. 5–12, 1991.

[54] V. Guruswami, *List decoding of error-correcting codes: Winning thesis of the 2002 ACM doctoral dissertation competition.* Springer Science & Business Media, 2004, vol. 3282.

[55] V. Guruswami, R. Li, J. Mosheiff, N. Resch, S. Silas, and M. Wootters, "Bounds for list-decoding and list-recovery of random linear codes," *IEEE Transactions on Information Theory*, vol. 68, no. 2, pp. 923–939, 2022.

[56] A. Rudra and M. Wootters, "Average-radius list-recoverability of random linear codes," in *Proceedings of the Twenty-Ninth Annual ACM-SIAM Symposium on Discrete Algorithms*, ser. SODA'18. USA: Society for Industrial and Applied Mathematics, 2018, pp. 644–662.

[57] A. V. Kuznetsov and B. S. Tsybakov, "Coding in a memory with defective cells," *Problemy Peredachi Informatsii*, vol. 10, no. 2, pp. 52–60, 1974.

[58] A. El Gamal and Y.-H. Kim, *Network information theory*. Cambridge University Press, 2011.

[59] A. D. Wyner, "Random packings and coverings of the unit n-sphere," *The Bell System Technical Journal*, vol. 46, no. 9, pp. 2111–2118, 1967.

[60] C. E. Shannon, "Probability of error for optimal codes in a Gaussian channel," *Bell System Technical Journal*, vol. 38, no. 3, pp. 611–656, 1959.

[61] J. Schalkwijk and T. Kailath, "A coding scheme for additive noise channels with feedback–I: No bandwidth constraint," *IEEE Transactions on Information Theory*, vol. 12, no. 2, pp. 172–182, 1966.

[62] R. G. Gallager and B. Nakiboğlu, "Variations on a theme by Schalkwijk and Kailath," *IEEE Transactions on Information Theory*, vol. 56, no. 1, pp. 6–17, 2010.

[63] I. Csiszár and J. Körner, *Information theory: Coding theorems for discrete memoryless systems*. Cambridge University Press, 2011.

[64] C. Berge, *Topological spaces: Including a treatment of multi-valued functions, vector spaces and convexity*. Oliver & Boyd, 1963.

About the Author

Yiming Yan (Chinese name: 闫一明) was born in Beijing, China on 17 February, 1996.

In 2014, she graduated from Beijing No.4 High School and joined Tsinghua University. There she studied Electronic Engineering and obtained a Bachelor of Engineering degree with distinction in 2018. During the summer of 2017, she did a three-month research project at the Institute for Communications Engineering (LNT) at Technical University of Munich.

In 2018, she went to Switzerland and continued her studies at ETH Zurich. She obtained a Master of Science degree in Electrical Engineering and Information Technology in 2020, and was awarded the ETH Medal and the Willi Studer Prize. Since 2020, she has been a Ph.D. candidate at the Signal and Information Processing Laboratory (ISI) under the supervision of Prof. Amos Lapidoth.

She married Yuhang Lu in 2021.

ETH Series in Information Theory and its Applications

edited by Amos Lapidoth

Hartung-Gorre Verlag Konstanz – http://www.hartung-gorre.de